Studies in the Psy<

Series Editors

Stephen Frosh
Department of Psychosocial Studies
Birkbeck, University of London
London, UK

Peter Redman
Faculty of Arts and Social Sciences
The Open University
Milton Keynes, UK

Wendy Hollway
Faculty of Arts and Social Sciences
The Open University
Milton Keynes, UK

Studies in the Psychosocial seeks to investigate the ways in which psychic and social processes demand to be understood as always implicated in each other, as mutually constitutive, co-produced, or abstracted levels of a single dialectical process. As such it can be understood as an interdisciplinary field in search of transdisciplinary objects of knowledge. Studies in the Psychosocial is also distinguished by its emphasis on affect, the irrational and unconscious processes, often, but not necessarily, understood psychoanalytically. Studies in the Psychosocial aims to foster the development of this field by publishing high quality and innovative monographs and edited collections. The series welcomes submissions from a range of theoretical perspectives and disciplinary orientations, including sociology, social and critical psychology, political science, postcolonial studies, feminist studies, queer studies, management and organization studies, cultural and media studies and psychoanalysis. However, in keeping with the inter- or transdisciplinary character of psychosocial analysis, books in the series will generally pass beyond their points of origin to generate concepts, understandings and forms of investigation that are distinctively psychosocial in character.

More information about this series at
http://www.palgrave.com/gp/series/14464

Paul Hoggett
Editor

Climate Psychology

On Indifference to Disaster

Editor
Paul Hoggett
University of the West of England (UWE)
Bristol, UK

Studies in the Psychosocial
ISBN 978-3-030-11740-5   ISBN 978-3-030-11741-2   (eBook)
https://doi.org/10.1007/978-3-030-11741-2

Library of Congress Control Number: 2018968353

Cover credit: Miemo Penttinen—miemo.net/gettyimages

This Palgrave Macmillan imprint is published by the registered company Springer Nature Switzerland AG
The registered company address is: Gewerbestrasse 11, 6330 Cham, Switzerland

Acknowledgements

Chapter 12 is a revised and updated version of Hoggett, P. and Randall, R. (2018) Engaging with Climate Change: Comparing the Cultures of Science and Activism. *Environmental Values*, 27, 223–243.

Contents

1   **Introduction**   1
    Paul Hoggett

Part I   Mostly Methods

2   **New Methods for Investigating New Dangers**   25
    Renee Lertzman

3   **Children and Climate Change: Exploring Children's
    Feelings About Climate Change Using Free Association
    Narrative Interview Methodology**   41
    Caroline Hickman

4   **An Integrative Methodology for Investigating Lived
    Experience and the Psychosocial Factors Influencing
    Environmental Cognition and Behaviour**   61
    Nadine Andrews

5   Emotional Work as a Necessity: A Psychosocial Analysis
     of Low-Carbon Energy Collaboration Stories   85
     Rosie Robison

6   Researching Climate Engagement: Collaborative
     Conversations and Consciousness Change   107
     Sally Gillespie

7   Climate Change, Social Dreaming and Art:
     Thinking the Unthinkable   129
     Julian Manley and Wendy Hollway

Part II   Mostly Findings

8   Emotions, Reflexivity and the Long Haul: What
     We Do About How We Feel About Climate Change   153
     Jo Hamilton

9   Leading with Nature in Mind   177
     Rembrandt Zegers

10   Attitudes to Climate Change in Some English Local
     Authorities: Varying Sense of Agency in Denial and Hope   195
     Gill Westcott

11   We Have to Talk About…Climate Change   217
     Robert Tollemache

12   Engaging with Climate Change: Comparing the
     Cultures of Science and Activism   239
     Rosemary Randall and Paul Hoggett

13   Conclusions   263
     Paul Hoggett

Index   267

Notes on Contributors

Dr. Nadine Andrews work is concerned with supporting individuals and organisations to live in more harmonious relationship with the natural world. She uses nature-based and mindfulness-based approaches to research, consultancy, coaching and facilitation.

Dr. Sally Gillespie practised as a Jungian psychotherapist before completing Ph.D. at Western Sydney University on the psychological terrain of climate change engagement. She lectures on ecopsychology at WSU. Her book *Climate Change and Consciousness: Re-Imagining Our World and Ourselves* (Routledge) will be published in 2019.

Jo Hamilton is currently a full-time Ph.D. student at the University of Reading, School of Geography and Environmental Science. She has a background in academic research on community energy projects, environmental action and facilitating groups for social change.

Caroline Hickman is currently working at the University of Bath and as a psychotherapist with children and young people following trauma. She qualified as an Integrative Psychosynthesis Psychotherapist with Revision and studied archetypal and cultural psychology with Thiasos. Currently researching children's relationships with nature and climate change.

Paul Hoggett is an Emeritus Professor of Social Policy at UWE, Bristol, where he co-founded the Centre for Psycho-Social Studies. He now works as a psychoanalytical psychotherapist and helped set up the Climate Psychology Alliance.

Wendy Hollway is an Emeritus Professor in Psychology (Open University). She works with psychoanalytically informed methodologies that go beyond consciously aware thought, in particular Social Dreaming, which is especially useful for researching climate change. She is a Board member of Cape Farewell, an organisation using art to raise awareness about climate change.

Dr. Renee Lertzman is a practitioner whose work bridges psychodynamic research and environmental issues. Working mostly in the USA but also in the UK she has developed trainings for those working in environmental change, scholars and researchers. She is author of *Environmental Melancholia: Psychoanalytic Dimensions of Engagement*, Routledge (2015).

Dr. Julian Manley is Social Innovation Manager at the University of Central Lancashire. He is on the Executive Committee of the Climate Psychology Alliance and Trustee for the Gordon Lawrence Foundation for the Promotion of Social Dreaming. He has recently published *Social Dreaming, Associative Thinking and Intensities of Affect* (2018).

Rosemary Randall is a psychoanalytic psychotherapist who helped develop the award-winning Carbon Conversations project. In 2015 she co-authored *In Time for Tomorrow: The Carbon Conversations Handbook* with Andy Brown.

Dr. Rosie Robison is an academic at the Global Sustainability Institute (Anglia Ruskin University). Her research focusses on sustainable consumption, energy policy and interdisciplinary working. Amongst other edited collections, in 2015 she coordinated *Behaviour Change from the Inside Out*, an open access book exploring psychosocial methods for sustainability.

Dr. Robert Tollemache has been a psychoanalytic psychotherapist for 30 years. His doctoral research examining climate change denial combined his interests in psychoanalytic thinking and climate change.

Dr. Gill Westcott research includes economic policies for sustainability and local authorities' action on carbon emissions. She has a background in counselling and psychotherapy, and helps manage a smallholding in Devon. She has helped found a number of social enterprises and is co-chair of Transition Exeter.

Rembrandt Zegers works on the boundary of culture and nature. His experience comes from the business, not-for profit and NGO world (Greenpeace). He lives in the Netherlands. Currently he is also part-time Ph.D. student at UWE, Bristol, researching several nature practices.

List of Figures

Fig. 4.1   Analysis procedure   69
Fig. 7.1   Caché. Detail from artwork by David Buckland
            and Dr. Deborah Iglesias-Rodriguez.
            www.bucklandart.com/art/cache-2012/   142
Fig. 12.1   Individual and social defences   256

List of Tables

Table 4.1   Data sources   68
Table 4.2   Exploratory coding in IPA   70
Table 4.3   Analysis focus   72
Table 8.1   A selection of Emotionally Reflexive Methodologies
             relevant to climate change engagement   156
Table 10.1   Views about climate change   202
Table 10.2   Outlook on the future. Q: I am contented
             and confident that those who are now young
             will inherit a good future   212

1

Introduction

Paul Hoggett

Asleep at the Wheel

Ice core data from Greenland and Antarctica has revealed the unique, perhaps unprecedented, climatic conditions which appeared on Earth approximately 11,700 years ago and which have provided the basis for agriculture, settled life and human civilization (Petit et al. 1999). It is these conditions, ones that marked the beginning of what geologists called the Holocene, that we are now systematically destroying (Lieberman and Gordon 2018). Climate change, soil and ocean exhaustion, and mass species extinctions are both symptoms of this destruction and speed its progress.

The impacts of climate change are already affecting us. Sea-level rise threatens the island states of the south Pacific and delta regions such as Bangladesh. Rising seawater temperatures mean that 67% of all shallow water coral in the Great Barrier Reef is already dead or dying.

P. Hoggett (✉)
Frenchay Campus, University of the West of England (UWE), Bristol, UK
e-mail: paul.hoggett@uwe.ac.uk

© The Author(s) 2019
P. Hoggett (ed.), *Climate Psychology*, Studies in the Psychosocial,
https://doi.org/10.1007/978-3-030-11741-2_1

Rising land temperatures threaten once arable land with desertification, the resulting displacement of peoples fueling conflicts in areas such as North Sudan and Syria. Changes in climate have what is called a 'multiplier effect' leading to changes in other systems such as social systems and soil systems.

We know the signs, the scientific evidence is now overwhelming with 97% of all peer-reviewed studies between 1991 and 2011 supporting the concept of human-caused climate change (Cook et al. 2013). The scientific community is also aware that these impacts do not occur incrementally, drip by drip in a gradualist way. Indeed some geologists are convinced that the transition into the Holocene itself was triggered by changes occurring within just a few years (Zalasiewicz and Williams 2012). Gradual changes in quantity can suddenly tip over into a change of quality, just as with a human being when steadily increasing levels of stress suddenly produce a breakdown creating an entirely new emotional state. Understanding the Earth as a dynamic system in which quantitative and qualitative changes interact has informed the notion of 'dangerous climate change' and the belief that beyond a two degrees increase in global average temperatures the possibility of sudden and uncontrollable systemic ruptures greatly increases.

For two decades attempts to bring governments together to tackle climate change were thwarted by North/South conflicts and national self-interestedness. Finally in Paris in 2015 all 195 of the world's nations overcame their differences and committed themselves to decarbonisation plans which would keep global average temperature rises to 'well below two degrees'. The very next year atmospheric concentrations of carbon dioxide breached the symbolic mark of 400 parts per million, something ice core data indicates last occurred over 800,000 years ago. The next year, the USA, the world's leading carbon emitter, declared its intention to pull out of the Paris agreement. Just as bad, but less widely publicized, many signatories to the Paris agreement such as the UK seemed to behave as if signing the agreement was equivalent to 'job done'. It looked as if it was then back to 'business as usual' with UK government pledges to support increased air travel (Heathrow's Third Runway) and a rowing back on state interventions which had, for example, nurtured the renewables sector.

According to the latest projections being prepared for the next Intergovernmental Panel on Climate Change, even allowing for no sudden ruptural change, a 'business as usual' scenario would lead to an increase in global average temperatures by a minimum of 3.0% by 2100, and if this was combined with a lurch towards competitive nationalism and the breakdown of the already poor international cooperation on issues such as climate change the projections are worse (Carbon Brief 2018). Let's face it, a three degrees increase would be dire (Lynas 2008). Large parts of the planet would be effectively uninhabitable, mass migration would make today's migration seem negligible and the impact on the oceans and agriculture is unimaginable. How we would prevent a descent into barbarism is difficult to know—many of the super-rich are already planning their escape (Rushkoff 2017).

What appears to be our collective equanimity in the face of this unprecedented risk is perhaps the greatest mystery of our age. Many have attempted to offer explanations. Some say that climate change is just too intangible, lacking concreteness and immediacy it haunts us as a hyperobject defying our abilities to get our heads around it. Others argue that as it has become politicized climate science has lost its ability to represent fact and truth and has become simply seen by one group as a political totem belonging to another group (Dunlap et al. 2016). Opinion survey data provide us with some insights. Drawing on data from over 100 countries a research team at Yale found that whilst 90% of those polled in North America, Europe and Japan were aware of climate change, compared to only 40% worldwide, they were much less likely to see it as a serious risk than those in developing countries (Lee et al. 2015). This chimes with a recent study of attitudes in the UK which draws on British Social Attitudes data (Fisher et al. 2017). Whilst over 90% of respondents in all age groups believed in the existence of climate change, only 36% saw climate change as being 'mainly' or 'entirely' due to human activity. As with the Yale study, level of education was the main predictor of attitudes, though the UK data suggest age is also an important factor—those least aware of and least concerned about climate change were older and less educated. The authors concluded that the majority in Britain appears to have fairly middling attitudes towards climate change. They know about it, and acknowledge a human component, but are overall relatively indifferent and apathetic about climate change. (Fisher et al. 2017, p. 23)

How can we deepen our understanding of this collective complacency, one that none of us in developed countries are immune to? This is the aim of the present volume of essays.

Psychology Lite

It's amazing to realize just how crude and simplified the models were that until recently guided political elites in their thinking about human psychology. Whether these elites had in mind the economic or political behaviour of citizens or, more specifically, their behaviour in relation to issues such as health, energy consumption or climate change, they essentially shared a perspective akin to Spock in Star Trek. This assumed that humans were logical creatures that sought to act in reasonable (e.g. self interested) and consistent ways on the basis of information that was available and relevant to the choices that faced them. Spockism was (and still is) a pre-psychological perspective, essentially one that draws upon neoclassical economics. In climate change it found expression in the continued belief in some policy quarters that information was the key to change, that once citizens had the right information, communicated in the right way, then the scales would fall from their eyes and they would adopt more sustainable lifestyles. This perspective was not one which just bedeviled climate change, in the world of public policy it found expression in 'public choice theory' which underlay attempts to introduce markets into the public sector (guiding theories about how parents 'choose' schools based on information such as league tables, etc.).

Gradually the limitations of this approach became clear and over the last two decades an alternative which stresses the limited or 'bounded' rationality of human beings has gained ground. The new fashion is for 'behavioural economics' which draws upon contemporary social psychology and advances in neuroscience to illustrate the ways in which human behaviour is more complex than Spockist accounts would have it. Indeed according to contemporary research it seems, to the dismay of Spockists around the world, that we are crowd following creatures who constantly use mental 'short cuts' and 'feeling' cues to act in ways that make us feel better about ourselves. These and other ideas (the messenger is often more important than the message, we often rely in an automaton-like way upon our 'defaults', etc.) have become the stock in trade of behavioural economists and thinkers such as Daniel Kahneman (2011) and Paul Slovic (2000). The emergence of neuroscience in recent decades has also offered a new perspective on the nature of less conscious influences on human behaviour. In their 2008 book *Nudge: Improving Decisions About Health, Wealth and Happiness* Richard Thaler and Cass Sunstein introduce the reader to the neuroscientific distinction between the conscious, reasoning part of the brain (which they call the Reflective System) and the part of the brain, the amygdala, sometimes referred to as the 'emotional brain', which responds quickly and subconsciously to a range of signals in our environment. They argue that the new approach has raised 'serious questions about the rationality of many judgements people make' (Thaler and Sunstein 2009, p. 7), in other words, it has drawn attention to the extent to which 'intentionality' had been overemphasized by Spockists.

Such ideas have had a big influence on new thinking about human failure to engage with climate change such as George Marshall's book *Don't Even Think About It.*

So far so good, but let's be clear, this doesn't mean that political elites are finally about to adopt a more profound understanding of the human psyche. Experimental social psychology and neuroscience still offer a fairly simple view of humankind and one which splits the individual off from society. Moreover, their methodologies are informed by a positivist view of science which means that they only study behaviour which can be subject to experimental control (controlled and replicable laboratory and/or questionnaire-based research, typically using students as research subjects). Drawing upon the natural rather than the social sciences this approach assumes that it is possible to study scientifically the relationships between what are called dependent and independent variables.

The dependent variable is typically some form of human behaviour, for example the frequency and duration of householders' use of the hot water shower, which is thought to vary according to a range of factors which are subject to experimental manipulation (and hence policy intervention) such as exposure to feedback (e.g. through the use of smart metres) about the energy one consumes when using one's shower. This kind of information is the independent variable. The relationship between dependent and independent variables is not direct but mediated by another set of factors (mediating variables) such as age and gender. It may be, for instance, that exposure to information on energy use has a positive impact for older women (reducing shower use) but an inverse impact for young men (increasing shower use).

Sophisticated models have now been developed in which a whole range of independent and mediating variables have been incorporated. A good example is *Mindspace: Influencing Behaviour Through Public Policy*, a recent Central Office of Information document on behaviour change.

Deep Psychology

However, the new approaches have a number of limitations that need to be overcome. The focus has been on linear, cause/effect relations (albeit affected by a range of mediating factors) that assume behaviour can be influenced (nudged) if not controlled. But positivism as a paradigm of control over nature, including human nature, is what brought us into this sorry mess of climate crisis in the first place. Complex systems operate in completely different ways to simple linear models, and the impossibility of control in complex systems is a founding belief of complexity theory. Rather than cause/effect relations, we need to think of relations which are mutually influencing, recursive, indeterminate, and where gradual changes in quantity can suddenly tip over into a qualitative shift.

Positivist models are concerned with the assumed relationship between cognition and behaviour. But cognition is not the same as meaning, it is a narrower concept which, unlike meaning, is split off from emotion which is often seen as cognition's antithesis. Indeed until very recently political science, public administration and policy studies scarcely gave emotion a mention (Clarke et al. 2006; Hoggett and Thompson 2012) and as a consequence policy making elites have been completely outmanoeuvered by the rise of populist movements who have grasped the power of emotion-laden connections to electorates' hearts (Westen 2007). Understanding the centrality of feelings to human experience is vital, particularly in an area such as human engagement with climate change. We give meaning to our world and our world in turn influences the meanings we give to it. These meanings are influenced both by our unique life history and by the social identities we acquire, identities which are both pregiven (e.g. class) and chosen (e.g. environmentalist, patriot, farmer). To have a social identity means to belong to an 'interpretative community' (Lorenzoni et al. 2007), that is a group of people 'like me' who tend to share common experiences of the world. And because we have multiple identities we belong to many different communities, each with their own meaning frames, frames which often rub up against and conflict with each other. Consider a white, male, Catholic garbage collector living in the USA. As a white working class man he may feel climate change is some kind of liberal hoax devised by an educated elite. But the Catholic in him may be aware of and stirred up by the Pope's Encyclical on the Environment and the responsibility of stewardship to God's creation. He takes pride in his job and is also troubled at times by the casual way in which people, particularly those in affluent suburbs, treat their mountains of waste. He has two young children and sometimes worries about what kind of world they will grow up in. Each of these identifications pull him in different ways.

Positivist models emphasize what can be seen and measured, hence the focus on behaviour. But much of what is most important to human experience is not visible and appears to go on inside our head in this thing we call 'the self'. There is undoubtedly an illusional quality to the idea that we have 'a' self. Indeed for much of the time different parts of the self (or different selves) appear to engage in conversation with each other. It is the summer of 2018 and a realistic part of us is aware that the heat is unprecedented and worries about whether this is what 'they' mean by climate change, and then another complacent part of us kicks in offering reassurance and telling us not to be so childish with our worries. It is through these kinds of 'internal conversations' that we negotiate our different social identities and their meaning frames and it is through these conversations that we manage our emotions, sometimes in creative and realistic ways and sometimes by distorting, avoiding or not attending to reality.

So rather than ignoring the self as a kind of irrelevant 'black box' as many positivists do or assume some kind of unitary self exists that chooses, thinks and acts 'as one' we need to grasp the self in all its complexity, as something dynamic, highly differentiated, relational and conflictual. Rather than a narrow focus on behaviour, climate psychology is concerned with the many ways in which we engage with climate change—how we avoid, deny, embrace or accept it, dream about it, get depressed, terrified or guilty about it, feel in two minds about it, think it is a sign of the Second Coming or homo sapiens' final and deserved comeuppance, ruminate about it, wake up at night because of it, can't get our head around it, feel that 'really climate change is such a drag', know that it is something one should worry about (even though one doesn't), and so on and so on.

Climate Psychology

To repeat what was said earlier, our collective equanimity in the face of the unprecedented risk posed by climate change is perhaps the greatest mystery of our age. Current psychological models can throw some light on this issue but they are insufficiently complex and too individualized to rise to the challenge that we currently face. We need a different psychology. Our richest psychological insights have come from literature, philosophy, world religions and the psychotherapies. From such sources we glimpse some of the complexity and mystery of the human. The raw passions that often dominate our thoughts and behaviours; the internal conflicts and competing voices that characterize our internal lives and give colour to our different senses of self; the effect of powerful outside forces on the way in which we think and feel about ourselves.

Viewed from this perspective it is possible to see how our attempts to defend ourselves against the feelings aroused by worsening climate change are mediated by deep-seated assumptions about ourselves and society. For example, a powerful sense of entitlement may help us to shrug off guilt and shame about our lifestyles, or a touching faith in progress can mitigate anxiety and induce complacency. Typically we will feel torn between different impulses, to face and avoid reality, between guilt and cynicism, between what is convenient for us and what is necessary for the common good.

Climate psychology draws upon a variety of sources that have been neglected by mainstream psychology including psychoanalysis, Jungian psychology, ecopsychology, chaos and complexity theory, continental philosophy, ecolinguistics and social theory. It attempts to offer a psycho-social perspective, one that can illuminate the complex two-way interaction between the personal and the political.

Climate change and environmental destruction threatens us with powerful feelings—loss, guilt, anxiety, shame, despair—that are difficult to bear and mobilize defences such as denial and distortion which can undermine our capacity to get to grips with the issue. Climate psychology seeks to understand how this plays out both in our individual lives and in society and culture.

Climate psychology seeks to further our understanding of:

- Collectively shared and organized feelings such as loss, despair, panic and guilt evoked in individuals, communities, nations and regions by climate change and environmental destruction.
- The defences and coping mechanisms, such as denial and rationalization, that we use to avoid facing these difficult feelings and how such defences have become integral to sustaining our exploitative relations with both the non-human and human worlds;
- The cultural assumptions and practices (e.g. the sense of privilege and entitlement, materialism and consumerism, the faith in progress) that inhibit effective change;
- The conflicts, dilemmas and paradoxes that individuals and groups face in negotiating change with family, friends, neighbours and colleagues;

- The psychological resources—resilience, courage, radical hope, new forms of imagination—that support change.

'Climate Psychology' has emerged in the last few years as a way of offering a more complete understanding of the non-rational dimensions of our collective paralysis in the face of worsening climate change. It does this by focusing on the role of the emotions and psychic defences as they are manifest in individual lives and group experience and by examining the mediating role of identities and discourses in shaping meanings and facilitating or inhibiting action. There is now a growing body of climate psychology literature which draws upon the variety of sources, listed above, that have been neglected by mainstream psychology (Rust and Totton 2012; Weintrobe 2013a; Dodds 2011; Randall and Brown 2015; Adams 2016; Orange 2017). The great majority of these authors are involved in the Climate Psychology Alliance (CPA). However with the exception of Lertzman (2015) there is an important gap in this literature—it is largely theoretical or draws upon clinical experience, but it is not based upon empirical research. This volume of essays seeks to rectify this omission.

Key Concepts

In this volume, the reader will repeatedly encounter a number of key ideas which inform both ways of thinking about the human condition and methodologies for researching it.

Psycho-Social

Most contributors to this volume have been influenced by the emergence of the field of psycho-social studies over the last two decades. Essentially a transdisciplinary approach to theory and research, psycho-social studies focuses on subjective experience in its social context. It is interested in exploring the constant two-way traffic between the private experience of the individual, their inner conversations, fantasies, dreams and feelings, and the world of family, work, leisure, culture, politics and, crucially for a book such as this, of nature. There are many debates within psycho-social studies including whether the two terms should be separated by a hyphen or not. The editor of this volume declares his belief in the value of the hyphen as something which suggests that whilst 'psycho' and 'social' are informed and constituted by one another, they are not reducible to one another. Moreover, the hyphen refers to the possibility of a third area, neither purely 'psycho' nor 'social', but which draws upon both (Hoggett 2008). In this volume 'the imaginal' (Hickman), the social dreaming matrix (Manley and Hollway) and social defences (Randall and Hoggett) are all examples of phenomena inhabiting this third space.

Unconscious

To say that something is not conscious means that it is unavailable for thought. Whilst we seek to give meaning to our experience by representing it in thought we may not be able to do this. Our experience may simply be too intangible to represent in this way, words and thoughts may just not do it justice. This is why we have art, music and poetry, to find ways of giving meaning to human experiences which in some way go beyond our capacity to fix them in words and thoughts. This is also the language of dreams, they have an allusive and elusive quality, there is always something unresolved about them, something that cannot be pinned down. We might think of such forms of experience as 'pre-symbolic'. And this is how we all started out in life as infants, inundated with experience of different kinds but with such a limited capacity to give representation to them. As we shall see later, climate psychology research seeks to devise methods of gaining access to such forms of experience by deploying imagery, metaphor, narrative and dreams in its research methodologies.

Something may also be unavailable for thought because it more or less escapes our notice. Many habitual and routinized forms of life are like this, we sometimes refer to such experience as 'tacit' or 'implicit'. And much of our decision-making is like this; rather than the outcome of conscious decision-making, it occurs 'automatically' or spontaneously. We jump in a car and go to the shops without giving it a thought. And at a deeper level, many of our meaning frames operate like this, beyond our notice. Take, for example, the ideology of progress and the assumptions that are a part of it—for example, that human ingenuity can overcome all problems—we are often completely unaware of just how pervasive such assumptions are in our everyday lives. Hence our shock and panic when parts of the technological infrastructure of life collapse and we are left without electricity to heat our homes, charge our mobile phones and bring food to our supermarket shelves. As climate breakdown proceeds such shocks will occur more frequently.

Finally, there is the traditional psychoanalytic conception of the unconscious, that something is unavailable for thought because we lack the capacity to contain the powerful feelings that it would evoke in us. In relation to climate change this refers to grief (Randall 2009; Head 2016; Lertzman 2015; Hoggett 2018), to guilt (Randall 2013) and shame (Orange 2017) as well as to anxiety and despair (Weintrobe 2013b). As a consequence, we shut such feelings out using defence mechanisms such as denial, rationalization, splitting and dissociation. By using these mechanisms we keep the thought separate from the feeling element of our experience so that, whilst we may know about climate change, in an odd kind of way it leaves us undisturbed, a mechanism called disavowal. We have first-hand experience of the heat and the dryness of the summer, we read about temperature records being broken in the news, we hear about the likelihood of food price rises as a result of poor harvests and yet we remain partially asleep, without urgency or motivation to do anything.

Affect and Emotion

The distinction between conscious and unconscious experience leads us to another important distinction, between emotion and affect. Many of the feelings such as grief that I have just mentioned are only aroused when someone becomes, to some degree, aware of climate change and it is in the nature of an emotion that it has a conscious object (climate change) and meanings (risk, loss) attached to it. But there is another kind of feeling, commonly referred to as an 'affect', which operates at a much less conscious level. Anxiety and depression (the core ingredients of stress) are classical examples. Typically when we are anxious our feeling flits from one thing to another, it has no secure object to attach itself to. It is also very visceral, felt primarily through the body rather than through cognition. It is important not to reify this distinction into a split between affects on the one hand and emotions on the other but rather to think of human feelings as a continuum with meaningful emotions such as grief, fear and guilt at one end and affects at the other. The distinction is nevertheless important because it enables us to understand how people, such as those in denial, may not be 'climate aware' and yet are nevertheless affected by powerful feelings provoked by climate change.

Containment

The distinction between emotion and affect is helpful for another reason. Generally speaking the more a feeling can be made conscious and therefore subject to reflection the more it can be contained. 'Containment' refers to the extent to which an experience can be digested and worked through or, put another way, the extent to which a feeling can be transformed into an emotion as opposed to remaining as an affect. The psychoanalyst Wilfred Bion (1962) likened the mind to a system for digesting experience. If an experience can be contained, even a difficult one, it will provide food for thought (and therefore for growth and development). But if it can't be contained then, like undigestible food, it gets stuck in the system or has to be got rid of. So if we can't contain feelings of anxiety they will either get stuck in our system (paralyzing despair) or we will get rid of them by projecting them into others (we find an object for our anxiety, an Other) or through blind forms of action (because action can itself be a way of getting rid of uncontainable feelings). Simply put, are we able to use our feelings or do they use us? Are we able to regulate and cope with them and use them creatively or are we going to be at their mercy, as is usually the case when we deploy rigid defences such as denial and splitting?

Containment is essentially provided by social support, someone to talk to, share experiences with and find meaning in what before had been unthinkable. When we have lost a loved one the support of another individual can help us in this way. But containment can also be provided by the group, by institutions, and by the art, literature, customs and rituals of a culture. In relation to climate change there is growing experimentation with methods of providing such containment, some of which is examined by Jo Hamilton in this volume.

Different Ways of Knowing

A final important theme running through this volume is that rational, ana-lytical thought is just one way of apprehending reality. Whilst this method is crucial to science, the humanities also offer another way of knowing, perhaps one that offers even greater insights in relation to human reality. Clearly, the social sciences are situated in between and yet the academic discipline of psychology, particularly in the UK and USA, has always located itself firmly within the sciences, often mimicking scientism as a way of gaining respectability. As Rembrandt Zegers (in this volume) notes, one of the problems with rationality lies in the way it reproduces a mind/body dualism which dismisses other ways of knowing (e.g. sensing, intuit-ing) that we share in common with our fellow but non-human creatures. The climate change community already recognizes that information- and cognition-based approaches to communication have a very limited effect and that different ways of engaging the public are necessary. This book gives other ways of knowing—aesthetic, imaginal, sensory, associational—their rightful place alongside cognitive and analytical ways.

Researching Human Engagement with Climate Change

I have argued that we need to shift the balance towards a psychology that is as concerned with experience, meaning and feeling as it is with behaviour and cognition. Considering our present collective paralysis in the face of climate breakdown we need to focus much more on what is less conscious and less easily put into words. We also need to get a better grasp of the complex interplay between what goes on inside individuals (inside the black box), the organizational and social environments that they participate in every day, and their experience of the non-human. It follows that we need different research methodologies to those common to positivist psychology. In short we need methods which are qualitative and conversational, focus on a small number of cases, seek to gain access to what is less conscious, that make use of imagery, metaphor, stories and dreams, and are reflexive and empowering. In the last two decades, such methodologies have begun to gain a foothold in the social sciences, particularly in the UK where they have formed part of the development of what has become known as psycho-social studies. From the outset it has been concerned to develop narrative-based methodologies which situate experience in terms of people's stories of both their early life histories and current social identities and moral/political/cultural attachments (Hollway and Jefferson 2000, 2013; Wengraf 2001; Clarke and Hoggett 2009). As it has developed it has incorporated a wider range of methods including the use of imagery, observation and ethnography (Cummins and Williams 2018).

Qualitative and Conversational

Mainstream psychology is quantitative, concerned with measuring relations between independent and dependent variables or with statistical analyses of variations within large populations (as with opinion survey methods). What it doesn't do is talk to people, to try and understand their experience. This is the purpose of qualitative methods.

A Small Number of Cases

Qualitative research provides case studies of individuals, groups and organizations. Cases are researched in detail and depth rather than in large numbers. Whereas quantitative research might tell us what is going on, qualitative research wants to understand why. So, to return to the introduction to this chapter, survey research tells us that approximately 90% of UK adults are aware of climate change but that the majority of those who are aware are also indifferent and apathetic towards it. Qualitative research seeks to understand the many reasons why this is the case and what better way of doing this is there than to talk to people.

The Art of a Good Conversation

To develop a dialogue with someone, particularly around a 'hot' issue like race or climate change, requires particular kinds of skills and qualities in a researcher. Thought must be given to how the interviewer intervenes, how to avoid leading questions or questions which close down rather than open up exploration. The interviewee is likely at times to feel embarrassed, guilty, anxious or angry and may react defensively, the interviewer needs to be able to recognize this and intervene in a way which can contain these kinds of feeling. Given the focus on experience, meaning and feeling the interviewer will encourage the elicitation of stories, memories, imagery and metaphor. They help 'open things up' so that the conversation develops an exploratory dimension.

The Reflexive Researcher

Qualitative research will always be to some extent 'subjective' because the researcher is given the freedom to generate a unique conversation, something that standardized, structured interviews try to eliminate. Rather than being defensive about this, psycho-social research seeks to make a virtue out of it. Researchers unquestionably will have their own agendas—they feel curious, passionate or disturbed about something, that is why they want to investigate it. Rather than hide this, a psycho-social approach strives for maximal self-awareness on behalf of the researcher—tuning into their own prejudices, projections and inability to listen to things they find uncomfortable to hear. This is what is meant by the 'reflexive researcher' (Bager-Charleson 2016; Holmes 2019).

Empowering Methodologies

A good research encounter with an individual or group generates a dialogue, at the end of which they will have the experience of being understood and recognized and, more often than not, the feeling that through the encounter they are able to say to themselves 'I had never thought of that before'. In other words, to use a concept introduced earlier, the encounter will be 'containing', one that leads to the digestion of an experience which has provided food for thought. Difficult feelings can be identified and shared in a non-defensive way with another who is trying their hardest to listen nonjudgementally and with respect and compassion.

References

Adams, M. (2016). *Ecological crisis, sustainability and the psychosocial subject: Beyond behaviour change*. London: Palgrave Macmillan.

Bager-Charleson, L. (Ed.). (2016). The creative use of self in research: Explorations of reflexivity and research relationships in psychotherapy. *The Psychotherapist: Special Issue, 62,* 1.

Bion, W. (1962). *Learning from experience*. London: Heinemann.

Carbon Brief. (2018, April 19). *Climate modelling explainer: How shared socio-economic pathways explore future climate change*. https://www.carbonbrief.org/explainer-how-shared-socioeconomic-pathways-explore-future-climate-change.

Central Office of Information. (2010). *Mindspace: Influencing behaviour through public policy*. http://www.instituteforgovernment.org.uk/sites/default/files/publications/MINDSPACE.pdf.

Clarke, S., & Hoggett, P. (Eds.). (2009). *Researching beneath the surface: Psycho-social research methods in practice*. London: Karnac.

Clarke, S., Hoggett, P., & Thompson, S. (2006). *Emotion, politics and society*. Basingstoke: Palgrave Macmillan.

Cook, J., Nuccitelli, D., Green, S., et al. (2013). Quantifying the consensus on anthropogenic global warming in the scientific literature. *Environmental Research Letters, 8*(2). http://iopscience.iop.org/article/10.1088/1748-9326/8/2/024024/meta.

Cummins, A.-M., & Williams, N. (Eds.). (2018). *Further researching beneath the surface, volume 2: Psycho-social research methods in practice.* London: Karnac.

Dodds, J. (2011). *Psychoanalysis and ecology at the edge of chaos.* London and New York: Routledge.

Dunlap, R. E., McCright, A. M., & Yarosh, J. H. (2016). The political divide on climate change: Partisan polarization widens in the U.S. *Environment: Science and Policy for Sustainable Development, 58*(5), 4–23. Available at: http://doi.org/10.1080/00139157.2016.1208995. Accessed 16 Sept 2018.

Fisher, S., Fitzgerald, R., & Poortinga, W. (2017). *Climate change: Social divisions in beliefs and behaviour.* http://bsa.natcen.ac.uk/media/39251/bsa35_climate_change.pdf. Accessed 16 Sept 2018.

Head, L. (2016). *Hope and grief in the anthropocene.* London and New York: Routledge.

Hoggett, P. (2008). What's in a hyphen? Reconstructing psycho-social studies. *Psychoanalysis, Culture and Society, 13*(4), 379–384.

Hoggett, P. (2018). On being surrounded by loss. *Psychoanalytic Psychotherapy, 32,* 243–246.

Hoggett, P., & Thompson, S. (Eds.). (2012). *Politics and the emotions.* London and New York: Continuum.

Hollway, W., & Jefferson, T. (2000). *Doing qualitative research differently: Free association, narrative & interview method.* London: Sage.

Hollway, W., & Jefferson, T. (2013). *Doing qualitative research differently: A psychosocial approach.* London: Sage.

Holmes, J. (2019). *A practical psychoanalytic guide to reflexive research.* London and New York: Routledge.

Kahneman, D. (2011). *Thinking, fast and slow.* New York: Farrar, Straus & Giroux.

Lee, T. M., Markowitz, E., Howe, P., Ko, C.-K., & Leiserowitz, A. (2015). Predictors of public climate change awareness and risk perception around the world. *Nature Climate Change, 5,* 1014–1020.

Lertzman, R. (2015). *Environmental melancholia: Psychoanalytic dimensions of engagement.* London and New York: Routledge.

Lieberman, B., & Gordon, E. (2018). *Climate change in human history: Prehistory to the present.* London and New York: Bloomsbury Academic.

Lorenzoni, I., Nicholson-Cole, S., & Whitmarsh, L. (2007). Barriers perceived to engaging with climate change among the UK public and their policy implications. *Global Environmental Change, 17*(3–4), 445–459.

Lynas, M. (2008). *Six degrees: Our future on a hotter planet.* London: Harper Perennial.

Marshall, G. (2014). *Don't even think about it: Why our brains are wired to ignore climate change.* New York: Bloomsbury.

Orange, D. (2017). *Climate crisis, psychoanalysis, & radical ethics.* London and New York: Routledge.

Petit, J., Jouzel, D., Raynaud, N., Barkov, J.-M., Barnola, I., et al. (1999). Climate and atmospheric history of the past 420,000 years from the Vostok ice core, Antarctica. *Nature, 399,* 429–436.

Randall, R. (2009). Loss and climate change: The cost of parallel narratives. *Ecopsychology, 3*(1), 118–129.

Randall, R. (2013). Great expectations: The psychodynamics of ecological debt. In S. Weintrobe (Ed.), *Engaging with climate change: Psychoanalytic and interdisciplinary perspectives.* London and New York: Routledge.

Randall, R., & Brown, A. (2015). *In time for tomorrow? The carbon conversations handbook.* Stirling: The Surefoot Effect.

Rushkoff, D. (2017, July 5). Survival of the richest. *Medium.* https://medium.com/s/futurehuman/survival-of-the-richest-9ef6cddd0cc1.

Rust, M.-J., & Totton, N. (Eds.). (2012). *Vital signs: Psychological responses to ecological crisis.* London: Karnac.

Slovic, P. (2000). *The perception of risk.* London: Earthscan.

Thaler, R., & Sunstein, C. (2009). *Nudge: Improving decisions about health, wealth & happiness.* Harmondsworth: Penguin.

Weintrobe, S. (2013a). *Engaging with climate change: Psychoanalytic and interdisciplinary perspectives.* London and New York: Routledge.

Weintrobe, S. (2013b). The difficult problem of anxiety in thinking about climate change. In S. Weintrobe (Ed.), *Engaging with climate change: Psychoanalytic and interdisciplinary perspectives.* London and New York: Routledge.

Wengraf, T. (2001). *Qualitative research interviewing.* London: Sage.

Westen, D. (2007). *The political brain: The role of emotion in deciding the fate of the nation.* Washington, DC: Public Affairs.

Zalasiewicz, J., & Williams, M. (2012). *The Goldilocks planet: The four billion year story of earth's climate.* Oxford: Oxford University Press.

Part I

Mostly Methods

The chapters featured in this part focus primarily upon the methods used to generate and analyse data, although each also offers the reader examples of the kinds of findings generated by such methods. All six contributors illustrate the use of various kinds of psycho-social methodologies to investigate the way in which people engage with climate change and environmental destruction.

Renee Lertzman highlights the process of what she calls 'deep listening', that is, an empathic and nonjudgemental research attitude which influences how we ask questions, listen and interpret what we think we are hearing. She suggests this approach is particularly appropriate where we are researching the experience of those whose views about climate change, at face value, are very different to our own. She illustrates her argument by drawing upon interviews with conservative Republicans undertaken by a client who approached Lertzman's consultancy practice. The brief extracts from an interview with Jill vividly illuminate the complex and often contradictory narratives that influenced Jill's experience. They also reveal how the apparently opposing views of climate sceptics and climate activists may share some important values in common, for example, the emphasis on personal responsibility and criticism of the 'me' culture.

Using some delightful extracts from an interview with twelve year old Frankie, Caroline Hickman furthers this exploration of 'deep listening' by highlighting the contribution of what in Chapter 1 was termed the researcher's reflexivity, that detailed process of honest reflection both in the moment to moment encounter of the interview and afterwards. Hickman also makes a convincing case for the use of imaginal forms of conversation with children, in her case the use of 'personification' as in 'if climate change was an animal what would it be and what would it say?' Hickman argues that all of us, children and adults, engage in two very different kinds of thinking, one linear and rational, the other more playful and imaginal which makes use of imagery, metaphor, symbols, feelings and intuitions. This important distinction underlines several other contributions to this volume.

The theme of metaphor is pursued in detail in Nadine Andrews' contribution where she draws upon the work of George Lakoff and Mark Johnson to provide a detailed exposition of how to undertake frame and metaphor analysis of in-depth interviews conducted using a phenomenological methodology (IPA). Andrews reveals some of the metaphors sustainability managers use when talking about how they deal with the feelings of frustration, anger and despair provoked by working in governmental institutions dominated by a 'business as usual' mentality. She argues that such negative feelings were seen as a threat to the competency needs of these managers and hence suppressed and, along with autonomy and relatedness, she sees these needs, and not more surface views and values, as the primary drivers of pro-environmental behaviour.

Rosie Robison's research also focuses on the emotion work of those engaged in leading sustainability partnerships which often required complex, multi-stakeholder collaboration. She observes that whilst the use of narrative and story-telling methodologies is becoming more familiar in this field it currently gives little attention to emotion, something she seeks to rectify in her chapter. She describes an approach to the analysis of the stories told by sustainability workers which highlights the emotions which are often implicit in what otherwise appear to be bland and technical narratives. She is also interested in what these narratives reveal of the social defences deployed by organisations manifested in the way in which they think, speak and organise. This is something which is also highlighted in the study of climate scientists by Randall and Hoggett featured in Chapter 12.

The theme of different ways of knowing, featured in Caroline Hickman's contribution, is taken up and developed in a novel direction by Julian Manley and Wendy Hollway who describe an event organised for artists, scientists and activists in the climate change field which combined artwork and dream work. They used a method, the Social Dreaming Matrix, which has become increasingly influential in the last two decades in which groups of individuals share dreams provoked by a public issue such as political conflict, social exclusion or climate change. In this case, the artwork acted as a further 'provocation' to the dream work. They use ideas from Deleuze and Guattari to help us to imagine 'the imaginal', something associative, rhizomatic, embracing the bizarre-seeming and sensing the affect. This, they suggest, is more like a humanities-based research paradigm than a social scientific one, in which our struggle with the facts of science is replaced by a surrender to an aesthetic experience appropriate to the impossible-to-grasp nature of climate change.

Sally Gillespie's research focused on those she called 'the canaries' of climate change, that is, those activists whose awareness of the issue means that they are already exposed and vulnerable to its toxic psychological effects. Gillespie argues persuasively that research on the experience of activists embroiled in this deeply systemic phenomenon should be participatory, solidaristic and action-oriented. Her research is based on the experiences of a group of activists who met for a year under her convenorship and who managed their own process of group inquiry into the moral, existential and psychological dilemmas their climate awareness presented to them. Like Manley and Hollway, Gillespie was struck by the powerful role of dreams as ways of knowing and, like Hickman, she used imagery (in this case a map) to personify key themes emerging from the group's deliberations. Gillespie also notes the benefits group members derived from engaging in this collaborative inquiry, particularly in terms of managing trauma, taking back projections and avoiding simplistic good/bad, us/them, polarities.

2

New Methods for Investigating New Dangers
On the Radical Practice of Listening

Renee Lertzman

I was in the backyard of a cottage in Point Reyes Station in the summer of 2014, when I received the call, watching heavy bumblebees hover around an exploding bush of lavender-blue daisies and silver leaves. It was one of those perfect West Marin afternoons, a quintessential California scene, only a few miles in either direction from beach, woods or the famous local creamery housed in a barn across the street. So it was a surreal moment when asked by a trusted member of the Republican Party, a messaging researcher, to advise on a project designed to convince the Republican GOP that climate is an issue their constituents care about.

The researcher explained a few more details. He runs a firm known for their use of a cutting edge technology—dial testing—to measure how messages resonate with audiences. He then advises on messaging about often charged, hot-button issues. As someone personally deeply concerned about climate change, he had thrown his energies behind launching this ambitious project, and was lining up interviews and focus groups with "climate

R. Lertzman (✉)
Fairfax, CA, USA
e-mail: Renee@reneelertzman.com

© The Author(s) 2019
P. Hoggett (ed.), *Climate Psychology*, Studies in the Psychosocial,
https://doi.org/10.1007/978-3-030-11741-2_2

skeptic" Republicans around the United States. The primary requirement was that I was to not speak about this project, and in exchange for our collaboration, I would receive the raw data, and be able to write about the research later—as long as I left his name out of it.

I responded yes without hesitation. Here I was, a California native, about as liberal as they come, standing barefoot on a deck under the warm summer sun, about to embark on a collaboration with a Republican researcher in Washington, DC, to produce research findings that would find their way to the GOP through a series of briefings. I couldn't imagine anything more exciting. Thus began what I nicknamed The Republican Climate Project. The target audiences for the project were moderates and conservative Republicans who ranged from slightly skeptical to very skeptical about the existence and severity of climate change. My client conducted a total of 28 in-depth, one-on-one interviews in four locations across the United States. Each interview lasted a total of 75 minutes. He then conducted two rounds of moment-to-moment dial testing on specific climate messaging, using scripts informed by the interviews, read by actors reflecting different points of view. The most successful tested script, scoring in the high 80s(100 = total agreement) was one we created together, with intentional and explicit acknowledgment of anxieties, ambivalence, and aspirations distilled and decoded through the interview content (what I call The Three A's, discussed below).

This essay presents a necessarily brief, partial, and high-level discussion of what I see as germane to a psycho-social approach to researching how we respond, feel about and engage with climate change. I draw on this project as an example of one project that applied psycho-social methodology with great impact and success. I believe such approaches reflected in this volume represent critically needed innovations in how we do our work as researchers, for the sake of ourselves and our planet.

Good Insights Need Good Methods

At a time when increasing attention is focused on understanding people's motivations, perceptions, and attitudes about climate, it behooves us to think very carefully about our research methods. This includes whether motivation, perceptions, and attitudes are what we should be paying attention to. For decades now, many of us trained in social sciences continue to design and use methods that are arguably inappropriate for the task of addressing our situation of climate change. What I've found over decades of research methods training, conducting my own research, and now training others in psycho-social research methods specific to environment and climate, that our capacity to truly unlock key insights starts with research design. And, when it comes to research design, our mind-set often revolves in well-worn grooves about how we ask questions, interpret data, and relate with our participants. These deep grooves, perpetuated in research methods courses around the world, have influenced how all kinds of sectors conduct their research, whether it's a governmental agency, a charity, or a market research firm. This research often is fixated on uncovering people's attitudes, perceptions, and motivations (Lertzman 2015). There are many problems with this lens, most significantly an implicit assumption that people *lack* something, be it motivation, care, or concern; and the key is to uncover the *barriers* that may magically unleash the kind of responses we so urgently need. This kind of thinking is so pervasive across so many sectors, that I have come to see this as having *colonized* the ways in which we understand the dilemma of climate and environmental action.

Despite admonishments from influential researcher-critics such as Elizabeth Shove (2010) to drop our obsession with "behaviour change" and the holy trinity of "attitude, behaviour, and choice (ABC)," persistent assumptions remain about how best to research, and therefore gain insight into our complex relationship with climate change. Given the supreme importance of having "good data," "metrics," or "insights" to shape and inform practice, from governmental or nongovernmental sustainable behavior campaigns to multimillion dollar marketing efforts to engage people to question their consumptive practices, this situation simply must be addressed and rectified. To do so, we must be willing, as I have argued elsewhere (Lertzman 2015), to recognize the limits of our research methodologies, and therefore assumptions about metrics, data, meaningful insight, and go about the task of skilling up, scaling capacities, and supporting innovative approaches to research methods. This involves an active engagement with the specific tools and offerings afforded by what's considered "psycho-social research" communities. This includes understanding affect, unconscious (or nonconscious) defense mechanisms, cognitive dissonance, anxiety, guilt, shame, things we understand intuitively, even if in our own private lives, but haven't yet figured out exactly how to engage with as climate researchers. Perhaps more significant, it requires modifying the simple practice of how we ask questions, listen, and interpret what we think we are hearing, based on insights in related fields in clinical psychology, public health, and neurosciences. This is what I see as the profound opportunity and capacity psycho-social research methods offer us.

When I undertook doctoral research training in the UK between 2005 and 2010, I was housed in a highly reputable school of social sciences, full of academic stars and recipients of competitive research funds. As I came to see, much of this work focused on sociology, policy, and science studies, with explicit tensions between the "quals" (qualitative researchers) and the "quants" (quantitative researchers) Thankfully, at that time, I was fortunate to start my coursework when a small but powerful group of psycho-social researchers were convening, a sort of enclave within the larger school of social scientists. This group of faculty included Valerie Walkerdine, whose innovative research methods—such as being vulnerable with participants, drawing on one's own experience in the data analysis, being exceptionally sensitive to dynamics in the research interaction—struck me as incredibly potent, powerful, and with tremendous promise if directed in the area of environmental threats. I recall reading the chapter "Working with Emotions" in *Growing Up Girl* (Walkerdine et al. 2001), now a standby I use in teaching and training courses, marveling at the bravery and creativity demonstrated in the research approach. I wondered, why are we not seeing more of this in the climate field, which is arguably so fraught, charged, psychologically and socially complicated? What does it look like to be a brave researcher in these times? What does research methodology based around deep listening look like? How can we show up in our research more fully human and compassionate—and how does this change what we think we know?

The Radical Act of Listening

There are numerous attributes of our current climate crisis that place it in a category, unlike any other social, political, or economic issue. I believe many of us are already familiar with the most obvious of these, such as the systemic nature of the problem, and the challenges this poses for representing the issue via language, imagery, and framing. This can lead to an excessive focus on making the issue more "tangible," immediate and personally relevant. Another attribute concerns the ways in which climate has become clearly politicized and positioned as an ideological issue (Haidt 2012), which is to say that specific industrial practices have come to stand for certain allegiances and alliances within political tribes. Within the psychological sectors, we have seen emerging work addressing emotional stakes of the climate crisis, whether articulated as "psychological barriers" (Stoknes 2015) or as relating to forms of loss and grief (Kiehl 2016; Randall 2009). More recently, attention has been given to the issue of scale, oftentimes translating to a "hope versus despair" discourse stemming from the overwhelm, helplessness, inefficacy, and powerlessness that many individuals experience when learning about climate change (Kiehl 2016; Lertzman 2015). Two notable publications—in the New York Magazine and the New York Times respectively—catalyzed a national debate in the United States about the role of hope and despair in how the climate story is told (Wallace-Wells 2017; Rich 2018).

However, underlying the discussions about what makes climate change so exceptionally difficult to respond to at an individual, social, political, or economic level are the ways in which the experience of climate change is often a "brew" of incredibly complicated feelings, attachments, drives and sentiments. The operative word here is *experience*. Whereas many of us who wish to understand the how and why of climate action, we are rarely asking the question about how the issue is experienced. I've found the frame of experience—versus opinion, attitude, belief, position, view, or even concern—to open our capacities as researchers and practitioners, to get at this complex mixture of anxieties, ambivalence, contradictions, and yes, aspirations and desires. When we are curious about experience, we are able to enter into a research design process that is potentially less reductive, and more nuanced. More importantly, it allows the people with whom we have the great fortune to engage, to have more freedom, space, and permission to show up more fully. Which is what we all seek as researchers, especially when the stakes are so high, to obtain meaningful "data" or "insights."

In sharing highlights from my work on The Republican Climate Project—which also reflects approaches used in dozens of applied research projects with organizations seeking to connect with their stakeholders, audiences, communities, and citizens more effectively—my hope is to illuminate and clarify what a psycho-social approach may look like, regardless of the scope and size of our study. I am deliberately choosing a practical study, i.e. outside of an academic context. I feel strongly that research both inside and outside of the academy shares a commitment (hopefully) to uncovering truth; and not to valorize academic research. Many practitioners currently have potential influence to engage hundreds, if not thousands and millions of people, whether through implementing a city-scale engagement program or a national marketing campaign; and having good, nuanced insights is key. I am confident many of these methods can be adapted, iterated on, and applied in numerous contexts. What matters perhaps most is the mindset and relationship we have toward our research and participants, one that is ideally informed by what I have come to call a "compassionate research attitude." This is an attitude of nonjudgment and empathy, as much as we can muster when the issues are so charged (I return more to this below). Much of what I brought to this project was based on my training and experience in psycho-social methods, informed by a number of influences, such as Biographical Narrative Interpretive Method (Wengraf 2001), Free Associative Narrative Interviews (Hollway and Jefferson 2000), the Psychoanalytic Research Interview (Cartwright 2004), and dozens of psycho-social and psychoanalytic researchers and clinicians too numerous to mention (see Lertzman 2012 for discussion of methodology; and Lertzman 2015). What these approaches share is recognizing that a safe, trusting atmosphere can support true connection and rapport; and that we can never take what is said at face value. The task is how we can decode and interpret this "brew" as best we can.

Approaching the project with my Republican client, I reflected on what would support his capacity to generate the most meaningful insights for his task. In other words, to explore at the deepest levels possible what his participants are experiencing when it comes to the issue of climate change; and, how to convert this into meaningful insights that can apply to designing a message and strategy. As a seasoned messaging researcher, the focus is usually on what specific terms, phrases, and words resonate with certain audiences and groups, and then decoding this into messages and frames, whether health, future generations, economic benefit, and so on. In this project, however, he wanted to do something different. To get "underneath" the obvious triggers and discourses to what their particular anxieties, ambivalences and aspirations are. For this, a radical approach to listening was needed.

The Three A's: A New Way to Listen

The Three A's (Lertzman 2015) are "code" for a framework I use in both research and communications practices focused on climate change and environmental issues:

Anxiety

Ambivalence

Aspiration

I created the Three A's as a simple framework to distil the psycho-social thinking and approach developed in my work on *environmental melancholia* (Lertzman 2015) that could be used by academic and non-academic researchers alike. The A's represent coexisting affective and experiential dimensions when it comes to climate change and ecological threats. The As often relate to tensions found in competing attachments, affiliations, identities, and desires (for example "I'm not an environmentalist, but I am deeply concerned about the declining fish populations in our local lake and want something to be done about it"). I have found dialing into these Three A's can help us in dropping the preoccupations with opinion and surface-level cognitive processes (what we think/believe), and focusing on what's less conscious, yet exerting tremendous influence—what we may be feeling or sensing on cognitive, affective, and somatic levels in our confrontation with our ecological crisis.

For the Republican Project, designing research interview protocol/ discussion guides and subsequent analysis and message scripting with the Three A's in mind meant thinking about the questions we asked, how we asked them, the tenor and mood of these questions, and whether to even call them questions. In the spirit of psycho-social research, I prefer "prompts" to "questions," as we are interested in having rich conversations or dialogues; prompts signal the potential for a free-associative mode of communicating. Rather than design prompts to elicit specific anxieties or ambivalence, we focused on how we can open up the topic in a safe, disarming, and nonconfrontational approach. This is something that many climate researchers, particularly those trained in the social and natural sciences, often overlook—these are incredibly complicated topics, affectively laden and charged, and eliciting defenses in our participants is counterproductive (Lertzman 2015; Hollway and Jefferson 2000; Cartwright 2004).

The way prompts are framed and designed borrows strongly from clinical contexts, including motivational interviewing practice (Miller and Rollnick 2012), also elaborated in the chapter "Why Methods Matter" (Lertzman 2015). The focus is on asking open-ended questions that elicit free-association, reflection, reverie, fantasy, and unguarded responses. The only way this is possible is by ensuring participants feel safe and assured their experience is not going to be evaluated, judged, or critiqued. Designing the "preamble"—how the interviews are introduced—is a key part of this. The researcher is encouraged to convey there are no right or wrong responses, transparency in the intention and goals of the interview and project, clarification of boundaries (i.e. the participant can terminate at any time, raise questions, and so on), and prepare the participant for the kind of conversational style of interaction. I find giving people context, such as stating, "I will be asking you some follow-up questions, and on the whole this is a conversation" can put people at ease. For the Republican Climate Project, we worked on designing prompts to uncover in a safe way the associations and feelings participants have about climate change, particularly what it brings up. I encouraged my client to go slow, not rush participants in responding, and to be sensitive to emotional cues and respond accordingly.

Designing research interviews in this way relies on a fundamental skill-set and attitude of deep, active, and present listening. It is the researcher's responsibility to ensure participants *feel* heard and acknowledged, and achieving this through practices of active listening and verbal and nonverbal cues. My client was able to demonstrate empathy throughout, in relating to what was shared when appropriate, and continually affirming that whatever was shared is valuable and meaningful. As a result, I believe we were able to generate far richer data, enabling us to create a successful script that acknowledged the Three A's, thereby establishing an empathetic connection with the audience.

Conflicted Narratives

In exploring anxieties, ambivalence and aspirations, we challenge the tendency to focus on values, beliefs and related "units" that emphasize consistency instead of inconsistency. Instead, in this approach, we are keenly interested in the places of contradiction, conflict, dilemmas, and tensions that can be conveyed through narratives. Rather than seeking continuity of values, we listen instead for the ruptures and repairs that signal areas of friction and irreconcilable tension. I have come to see, in overseeing or conducting hundreds of research interviews on topics concerning sustainability—whether it's transportation, food, waste, energy efficiency, or water—that a majority of people from varied backgrounds experience in some form or another a genre of "double bind" unique to the climate context (Bateson et al. 1963; Lertzman 2015). This has many variations and expressions, but the basic gist is the experience of "damned if you do, damned if you don't" or being "between a rock and a hard place." This shows up commonly in interviews and focus group discussions as the ability to both acknowledge the issues, even if indirectly, while stating the impossibility of any meaningful form of response. Often, the impossibility either relates to how response or solutions may conflict with one's identity (the "I'm not an environmentalist" position), or seem insignificant in light of the scale of the issues ("what can I do as just one person? Not much").

The Republican Climate Project interviews were no exception. Listening to the interviews through the filter of anxiety, ambivalence, and aspiration helped us uncover key themes appearing across many of them. We were able to decode certain aspects of contradiction to uncover the Three A's. The following excerpts are from a female participant in Florida; we will call her "Jill."

> Having the smog [in L.A.] is better than having the EPA get rid of it, because I think people have a choice, and if they're going to choose to live in that environment, I think it's their right to choose that. I don't think our companies can just put harmful things in the air. I don't think that the EPA would allow that, and the EPA should not allow it… *It's hard because the core of me wants as little government control and invasion in our lives as possible.* However, if a company has some type of manufacturing, I don't think that they should be able to produce something that would cause cancer, for instance. *But that kind of goes against my core of less government.* I haven't had to reconcile something like that personally, but I suppose that I would research this issue more and my voting choice could be impacted because it's important to have the right leadership.

We can see in this excerpt an apparent dilemma. While Jill's "core … wants as little government control and invasion in our lives," she also doesn't feel companies should be allowed to put pollutants into the air. Boxed into this ideological corner, she even seems to feel that people should have the "right" to breathe polluted air, which doesn't make rational sense. This dilemma appears to put her into what may be an irreconcilable "double bind"—less government, yet more protection and regulation. It leads to an incoherent position, which on the face of it can appear as self-destructive or undermining. A common lament within more liberal political circles tends to be, "How can people vote against their own self-interest? Don't they see that giving industries a free pass has direct impact on health and well-being?" The same goes for climate solutions, with many climate advocates puzzling over the refusal to endorse innovations and solutions which "clearly" (to them) have obvious benefits and serve all of us. However, in ignoring anxieties and ambivalence underlying aspirations—to be healthy, free and happy, for example—we routinely bypass the investments that keep many of us stuck and resistant to change. The experience is as an unsolvable problem. As Ramachandran wrote about this tendency to preserve what is known, even if it requires a refutation of reality,

> … The coping strategies of the two hemispheres are fundamentally different. The left hemisphere's job is to create a belief system or model and to fold new experiences into that belief system. If confronted with some new information that doesn't fit the model, it relies on Freudian defense mechanisms to deny, repress or confabulate—anything to preserve the status quo. (1999, 136)

Throughout the interviews, my client and I were able to listen both frontally and laterally, between the lines, of where participants were experiencing such tensions between their anxiety, ambivalence, and aspirations. What we found sometimes surprised us, but was consistent throughout: people appeared entirely capable of holding competing and contradicting feelings about these issues of climate, energy, conservation, and environment. In other words, the relationship with these issues was marked by ambivalence.

Another example from Jill is illustrative of such conflicts. In discussing where we get our energy from, she responds:

> I have no problem with us drilling and trying to find our own sources of energy domestically. Obviously, it would be great to use the least amount of resources possible, but I think realistically, we should be independent of other countries on our oil and our energy sources. Besides creating some kind of car that doesn't require gas, which I just don't see realistically as happening any time soon. I don't know… I don't think we can live in a community where we just power around in golf carts these days. Airplane fuel and all that kind of stuff, I don't know of any realistic alternatives for that kind of thing. In terms of how Americans use energy day to day, I said earlier that we should use as little energy as possible. If everyone were just responsible on their own and could just be mindful of their usage—water at the home, using energy efficient appliances. It's important to do that because I believe that there is a limited amount of

these resources. *I think we do need to be responsible for our future gener-ations and not just have this "me" mentality, where I could just take what I want without any consequences down the road.* Also, I don't believe that the government should tell you what to do about these things.

I believe a lot in personal responsibility.

If we think for the moment in terms of the Three A's, we can hear anx-iety concerning loss of personal autonomy and control. It's evident Jill believes in personal responsibility, and is anxious about further govern-mental "invasion." That said, she also clearly yearns (aspires) for a world where "everyone were just responsible on their own," mindful and respon-sible about water and energy usage, and "be responsible for our future generations." Considering the heavy focus on a "future generation" trope in climate communications circles, we can see in this excerpt how such a narrative frame would fall flat with someone like Jill, unless there was acknowledgment of the anxieties about losing control, and ambivalence about wanting more people to be responsible (for a better world), yet not wanting more governmental control. The recourse is to fall back on a focus on "personal responsibility," a stance that flummoxes many climate advo-cates who recognize the profound need for increased levels of regulation and policy to support more feasible measures for scaled carbon reduction.

Acknowledging the Three A's

Where the rubber meets the road in our capacities to implement effective psycho-social research, is in being able to apply and imple-ment insights into tangible applications. This is where the Republican Climate Project became exciting. Our experiment was to pull out the themes of anxiety, ambivalence and aspiration, and see if we could create a script to acknowledge them and test the response. Clearly, we could not address each individual's "A's"—however, this pilot demon-strates that we don't need to get too granular to be effective. We were able to clarify specific themes, such as fear of more government, feeling vulnerable, not being included in the important conversations, patriotic pride, desire for doing things "on our own terms," a deep and abiding appreciation and love of the great outdoors, and core family values. We even included a reference to Al Gore, as someone who triggers immense anger and resentment about perceived hypocrisy of "flying around the globe, telling people to do what he's not willing to do himself," i.e. reducing carbon footprint.

As I am not at liberty to reproduce the script we created, I can highlight the strategy used for designing the script. We decided to explicitly speak and reference the various "A's" heard throughout many of the interviews, as a way to disarm and establish rapport with our conservative audiences. We began with a conservative-centric narrative about what "we" love most about being American—wide open spaces, freedom and ability to get away from it all, and cherishing nature and family. We then introduced the topic of climate change, as "changes to our climate," and the range of resentments, anger, and reactions that may come up. We included phrases like "you may feel angry or resentful…" and addressed the fear of having solutions handled by those who may not care about "your" best interests. The turn, however, comes when the narrator asks the audience, "What if you were to learn we can address this on our own terms?" Not surprisingly, the response rate soars at this point. The narrative then goes on to tap into deep aspirations about hard work, ingenuity and the qualities we are known for when it comes to tackling massive problems and innovations.

The result was a script—an actual script read by actors, tested with dial testing methodology—about climate change that truly resonated with a surprising number of conservatives who are skeptical about climate change issues. It was a powerful testament to the veracity of conducting research with a focus on radical listening, attuning to the Three A's, and acknowledging these explicitly and creatively through communications design.

Final Thoughts

When researching climate change, energy and environment—and the related policies, practices, and behaviors attending these massive issues—we must handle our research design with utmost care. By this, I refer to recognizing the affective dimensions of these issues, as topics that will activate a complex "brew" of guilt, shame, conflict, remorse, loss, grief, and so on. For many people, particularly in industrialized regions or the "Global North," these responses may be inchoate and felt, rather than thought, or what Bollas calls "the unthought known" (Bollas 1987). This poses a profound implication for anyone conducting research in these fields, as we also now know more about how affect (including cortisol, and related physiological processes) directly inform our capacities for processing information, learning, and accessing our own volition, feelings and perceptions (Alcorn 2013; Ramachandran 1999; Cacioppo et al. 2002). I maintain that climate, energy, and environment are domains that require new research methodologies that can explore and account for the ways in which we relate to industrial practices as both damaging and, at least at present, often supporting our lifestyles, identities, and ways of being. As researchers, we are responsible for conducting our investigations informed by established insights in clinical and neuroscientific studies, that offer tools for working with defense mechanisms, trauma, and unconscious processes. The new dangers of our time require new methodologies, of which psycho-social research is a key contributor and resource. This said, what is urgently needed are opportunities and resources for accessing capacity building and training across academic research initiatives and programs, as well as in communities of practice across government, nongovernmental organization (NGO), private, and public sectors. We need more brave and innovative researchers, willing to test, pilot, and prototype new methodologies for understanding our responses to a changing planet. We need active partnerships and collaborations across our disciplinary worlds. We need more people like my Republican client, who was willing to try out new approaches, and develop new capacities, so we can be better equipped to ideally support greater numbers of people in accessing our own innate capacities for creativity, imagination, resilience, and foresight.

References

Alcorn, M. (2013). *Resistance to learning: Overcoming the desire not to know in classroom teaching*. New York: Palgrave Macmillan.

Bateson, G., Jackson, D., Haley, J., & Weakland, J. (1963). A note on the double bind. *Family Process, 2,* 154–161.

Bollas, C. (1987). *The shadow of the object: Psychoanalysis of the unthought known.* New York: Columbia University Press.

Cacioppo, J. T., et al. (2002). Do lonely days invade the nights? Potential social modulation of sleep efficiency. *Psychological Science, 13,* 384–387.

Cartwright, D. (2004). The psychoanalytic research interview: Preliminary suggestions. *The Journal of the American Psychoanalytic Association, 52,* 209–242.

Haidt, J. (2012). *The righteous mind: Why good people are divided by politics and religion.* New York: Pantheon.

Hollway, W., & Jefferson, T. (2000). *Doing qualitative research differently: Free association, narrative and the interview method.* London: Sage.

Kiehl, J. (2016). *Facing climate change: An integrated path to the future.* New York: Columbia University Press.

Lertzman, R. (2012). Researching psychic dimensions of ecological degradation: Notes from the field. *Psychoanalysis, Culture and Society, 17*(1), 92–101.

Lertzman, R. (2015). *Environmental melancholia: Psychoanalytic dimensions of engagement.* London: Routledge.

Miller, W., & Rollnick, S. (2012). *Motivational interviewing: Helping people change* (3rd ed.). New York: Guilford Press.

Ramachandran, V. S. (1999). *Phantoms in the brain: Probing the mysteries of the human mind.* New York: Harper Books.

Randall, R. (2009). Loss and climate change: The cost of parallel narratives. *Ecopsychology, 3*(1), 118–129.

Rich, N. (2018, August 1). Losing earth: The decade we almost stopped climate change. *New York Times Magazine.* https://www.nytimes.com/interactive/2018/08/01/magazine/climate-change-losing-earth.html.

Shove, E. (2010). Beyond the ABC: Climate change policy and theories of social change. *Environment and Planning A, 42*(6), 1273–1285.

Stoknes, P. E. (2015). *What we think about when we try not to think about climate change.* Vermont: Chelsea Green Publishing.

Walkerdine, V., Lucey, H., & Melody, J. (2001). *Growing up girl: Psycho-social explorations of gender and class.* London: Palgrave Macmillan.

Wallace-Wells, D. (2017, July). The uninhabitable earth. *New York Magazine.* http://nymag.com/daily/intelligencer/2017/07/climate-change-earth-too-hot-for-humans.html.

Wengraf, T. (2001). *Qualitative research interviewing: Biographic narrative and semi-structured methods.* London: Sage.

3

Children and Climate Change: Exploring Children's Feelings About Climate Change Using Free Association Narrative Interview Methodology

Caroline Hickman

Interviewer question – 'Can you tell me how you feel about climate change?'

Participant answer - 'Aaaaarrrggghhhhhh'

Introduction

Most research into feelings and attitudes to climate change to date has been conducted with adult participants, not children; 'children are largely left out of discussions about appropriate responses to climate change, but they ought to be central to these debates because they, as well as future generations – have a much larger stake in the outcome than we do' (Currie and Deschênes 2016, p. 4).

C. Hickman (✉)
University of Bath, Bath, UK
e-mail: C.L.Hickman@bath.ac.uk

© The Author(s) 2019
P. Hoggett (ed.), *Climate Psychology*, Studies in the Psychosocial,
https://doi.org/10.1007/978-3-030-11741-2_3

This chapter will look at how the Free Association Narrative Interview method (FANI) developed by Hollway and Jefferson (2013) was used to research children's feelings about climate change. Influenced by Romanyshyn (2013) it also discusses how researchers using this methodology can 'play' with the imaginal landscape of the work engaging children in research about climate change using personification and metaphor. The FANI method asks us to examine both conscious and unconscious feelings, and as unconscious communication is often metaphorical in nature, the methodology also needs to embrace symbolic communication and data. The chapter concludes with extracts and reflection from interviews.

Climate Change Research with Children

So why research children's feelings about climate change? Research with children can open up new debates both for children, but also for society more generally (Tisdall et al. 2009), it can bring issues into light that might not otherwise have been considered, and offer new options for future action. I wondered if perhaps the way that children express things and see things could help adults think about their own feelings and responses to climate change.

Facing the challenges posed by climate change is a critical issue for us all (Clayton et al. 2014; Currie and Deschênes 2016), often framed in reference to the likely impact on children and future generations (Ojala 2012; Xu et al. 2012) with a focus on the impact on children's health and well-being (Stanley and Farrant 2015) and on their human rights (Gibbons 2014). The effects of climate change on children are expected to be considerable, particularly in the area of well-being and health, most profoundly affecting children in developing countries where social inequality is more pronounced.

Given this, I wanted to use a research methodology that would allow a range of children to participate in this research, children of different ages and cultural backgrounds; and I wanted to try to find out how they felt, not what they learnt at school, from TV, books or parents. I also wanted a methodology that could be used with children based in the UK, and also in countries that were facing the immediate impact of climate change who did not have English as their first language. I found that FANI as a methodology lends itself to research with children of different ages and backgrounds because it can be used playfully, even when talking about sensitive subjects, so it is appropriate for younger research participants and addresses many of the ethical concerns about researching with younger children (Phelan and Kinsella 2013; Tisdall et al. 2009; Darbyshire et al. 2005).

Research Design

The FANI method developed by Hollway and Jefferson (2013) was used to conduct this research using open-ended narrative questions aiming to collect data less conditioned by familiar discourse and also perhaps to capture internal conflicts and underlying attitudes not easily accessed consciously. FANI particularly suits research with defended subjects, and although children generally might not fall into this category I considered that climate change was a subject that people defend against. As a methodology used with children it allowed the child to set their own agenda and gives freedom to the interviewer to follow different narratives during data collection. Whilst it could be argued that children could be a less defended group than adult participants in research, they are also strongly influenced by prevailing social attitudes and by parental and family attitudes; so whilst not individually as defended as adults, I wanted to allow for the fact that children may mirror a general 'defendedness' that exists about the subject of climate change.

The research design is also based on an imaginal approach to research, using processes that deliberately acknowledge the dynamic field between the researcher and the research, and between researcher and participants. Through reflective analysis and interpretation this dynamic can be made as conscious as possible during data collection and in later interpretation through reflection. The researcher should 'play' with the 'imaginal landscape' (Romanyshyn 2013, p. 145) of the work, engage in transference dialogues, use active imagination, let go of the work (or at least try to let go of the ego attached to the work), use a reflective process of reverie to establish dialogue between the conscious and unconscious mind of the researcher, and stay uncertain.

This narrative inquiry also allowed for dialogue between researcher and participant that is open to change throughout the interview, with the intention being not to guide or suppress the participants' stories, but to encourage open and unstructured dialogue. The role of the researcher is to pay attention to the story and the storytelling, rather than to structure it around a set of questions and encourage the human 'impulse to narrate' (White 1980, p. 5). Subjective meaning can be explored as the stories or narratives emerge.

Free Association and Narrative

By using the method of Free Association in this research (encouraging the research participants to say whatever comes to mind) my intention was to try to make a space for a narrative not structured or based on conscious logical thinking, but based on unconscious process. 'The associations follow pathways defined by emotional motivations, rather than rational intentions' (Hollway and Jefferson 2013, p. 34). The narrative developed using the Free Association method therefore has a different approach to coherence; with incoherence, incongruence and defensive communication important to note and include in the data. In fact, the messages from these can be more telling and offer more important data than the conscious responses to questions. These can then be interpreted within a psychoanalytic theoretical framework, giving the coherence and containment needed for analysis.

In practice Holmes (2013) identified the following as guidelines for free association in research: asking open questions, thinking about answers beyond their face value, following participants dialogue rather than having a list of pre-determined questions, trying to ensure minimal intervention by the researcher, responding in non-verbal ways, and not closing off uncomfortable or anxiety provoking or irrelevant seeming narrative.

Intersubjectivity and the Research Relationship

Using FANI methods the researcher's role is not limited to that of eliciting and encouraging narratives. The narratives need to be understood
within the relationship developed between researcher and research participant, as both share the research field. The data is developed between
them with the feelings and thoughts of the researcher also informing
the data collected. Given this, both conscious and unconscious material from researcher and research participant should be explored and
noted in collection of data (Hollway and Jefferson 2013; Clarke and
Hoggett 2009). However trying to take into account the researcher's
own unconscious is of course a complex task, and whilst we cannot
be sure of the end results, it is 'worth the risk, because without any
systematic attempt to assess the presence of unconscious dynamics
in research, these dynamics function anyway in an invisible manner'
(Romanyshyn 2013, p. 105).

Researching the Unconscious

So, how should researchers 'dream the data' (Romanyshyn 2010, p. 4)
with communication from the unconscious never literal but symbolic,
how can we set aside any need for certainty and allow the meaning of
the symbol to show itself to us to try to access unconscious material?
This argument leans heavily on archetypal psychology's argument about
the soul of research (Hillman 1975) appearing in unexpected ways, and
never when we are looking directly at it, even less so if looking for it. It
means having 'an eye for the invisible in the visible, or using that third
ear that is sensitive to the whispers of those unanswered questions that
make their appeal to us' (Romanyshyn 2010, p. 85). It means surren-
dering the personal ego to the work of the research, paying attention
to the research that wants to be done, rather than the research that the
researcher wants to do.

Research Questions

Rather than just ask the single question 'how do you feel about climate change?', I felt the need to have some introductory questions with the children to build a relationship and start to 'play' with the research before asking this question. I also thought that in order to make the complex and sometimes intangible subject of climate change meaningful for children to engage with I would ask children questions about how they thought climate change might impact on the natural world before asking them how they felt about it themselves. My rationale for this was based on wanting to ask the children about something they may be able to immediately relate to. Climate change is a large, hard to grasp and a somewhat nebulous construct (see the chapter by Manley and Hollway in this volume); it can be difficult to relate to and has uncertain and contested impact on daily life for many. The majority of people in the West have largely ignored it, or engaged with it intellectually or abstractly; and it has been easy for us to defend against, or at its most extreme, to deny it (Lertzman 2015; Hamilton 2013; Weintrobe 2013).

The interrelationship between climate change and nature has been considered by a number of writers (Weintrobe 2013; Frantz et al. 2005) but conducted with adult participants. Research with children has examined conservation of different species and relationship with zoo animals and climate change (Clayton et al. 2014), but not explored the interrelationship between climate change and nature with children, nor examined children's own attitudes or feelings.

I developed the series of questions below, but used them loosely (and as you will see from the research extract at the end of this chapter, the children also had their own ideas, so we veered off course, or onto course whichever way you might look at it, almost from the very start!). Questions 1–3 were aimed at gathering data from conscious awareness; questions 4–6 were aimed at gathering more unconscious data using personification of climate change through an animal, bird or plant. Question 7 was used to reorientate the child back to the 'here and now'.

Narrative Questions in Individual interviews

1. Can you tell me about your thoughts and feelings about climate change?
2. Can you tell me about an animal or bird or plant that you like?
3. Can you tell me about the impact of climate change on that animal, bird or plant?
4. If climate change were an animal, bird or plant, what would it be?
5. If that animal, bird or plant could speak, what would it say about climate change, and how would it be feeling about climate change?
6. Do you have a message for adults about climate change?
7. What do you think about the conversation we have just had?

Personification, Metaphor and Active Imagination

Personification is the attribution of a personal nature or human characteristics to something that is non-human, or representation of an abstract quality into human form, and has been argued to be part of being human (Bering 2006; Kwan and Fiske 2008). It is an embodiment or symbol, image or representation of an object or thing in human form, which can help us to move into a relationship with the object and develop deeper meaning through bringing inanimate things to life Stewart (1998). Often used in literature, poetry and psychotherapy it includes attributes of feelings, metaphoric speech, personality and character 'as if' human.

Jung (1960) identified two kinds of thinking; linear and logical rational thought, and also an alternative way of thinking that relies on dreams, metaphor, art and ritual. Both are sources of knowledge, but of different kinds. Active imagination is a psychotherapy technique used to amplify images that emerge in the process of psychotherapy (Jung 1960; Schaverien 2005). I am interested in research as a relational process using active imagination and metaphor that can 'mobilise the psyche' (Schaverien 2005, p. 127) and so enable the participants to respond in ways that may not be entirely linear or rational, but are imaginal and maybe access unconscious thoughts and feelings.

Data Analysis

Analysis of data was conducted using psychoanalytically informed interpretative analysis and Romanyshyn's (2013) arguments concerning unconscious dynamics in the research process; not just examining the knowledge gained from the data, but also examining the researcher's relationship with the research process, or surface and depth phenomena Cooper (2009). In particular, the transference and countertransference responses were included as part of the data analysis. Psychoanalytic ethnography was used to explore how this unconscious projection plays a part in construction of our present day reality and perception of others (Clarke and Hoggett 2009). Interpretation was an important part of the research analysis with a focus on the emotional content of the research participants' experiences (Strømme et al. 2010) and with the researcher's own subjectivity used as part of the analysis in a reflexive way (Hollway and Jefferson 2013; Alvesson and Skoldberg 2000).

Interviews

To date I have interviewed 24 children, 8 in the UK and 16 in the Maldives, with research interviews due to be completed during 2019. Extracts from one research interview in the UK are detailed below showing the research process, followed by extracts from a number of other interviews. Interviews are on-going and analysis is incomplete at the time of writing. All the children were interviewed at their homes; with interviews varying in length, but averaging 20–30 minutes, the children interviewed ranged from age 5–17 including boys and girls, from rural and urban settings.

What follows is an interview extract with Frankie (12-year-old girl, UK) whose responses to me are in italics. This was the first interview I conducted using this methodology, and you can see my own uncertainty about using this method in my reflections. I have included some reflections from during the interview and some from later on when writing up the transcript.

CH - Can you tell me about your thoughts and feelings about climate change?

Frankie - 'Aaaaarrrggghhh'...........

CH reflection during interview—I'm aware that I'm uncomfortable with this as an opening question. It still risks sounding to me as if there is a right or wrong answer.

CH reflection after—Or maybe I'm struggling with the relationship between structure and open questions in this methodology. Maybe that's my transference—maybe I have an idea about what a 'researcher' should sound like and I'm trying to be that when actually another part of me just wants to use the single question approach? My countertransference asks me, is she feeling stupid or inarticulate? Am I?

CH – Oh (I laughed in surprise slightly), say more......

Whilst saying '*aaarrrggghhh*' she throws her hands up in the air, looking at me, and then away, she half laughs, but looks a bit scared, she glances at me and then back as if looking for reassurance. I wonder if she thinks she's answered wrongly because she's made a noise rather than said something that sounds more like a 'proper' answer. I just wait and smile and nod at her.

CH reflection during interview—I love her reply to be honest, it's not what I expected, but it's emotional and expressive and playful, and I love that she makes a noise rather than tries to answer more 'sensibly', with words.

CH reflection after—I don't think an adult is as likely to make noises in reply to a question. This was a response that was wonderfully expressive and open to interpretation! I was worried about that as an opening question, I wondered if it was too direct to go straight to that question, or might be off-putting for children to be asked that so directly. I now also wonder about my reassurance as well, is it reassurance about the research process that I wanted, or about climate change, from this child? Was I feeling some guilt about being an adult who has power (relative) to do something about climate change but am failing to do so (as are all adults). Was I reminded of this by this child? I know that I often feel guilt about climate change in relation to children and wonder what they will say to us in later years when there is so much less that we can do to mitigate the changes needed.

> *Frankie - 'I did it because..........and it's not enough, it's hard to explain how I feel and what I think.........'*

She pauses and seems to think about her initial answer. I wait and just smile at her.

CH reflection after—I liked that she says that *'it's hard to explain how I feel and what I think'*. Her willingness to show her uncertainty suggested to me that she was comfortable and relaxed enough to 'not know' or try to come up with a prepared answer, or make something up just to 'get things right' and come up with something more concrete. I think it's ironic, that I needed reassurance about my anxieties, and there it is, she is showing me how to stay with uncertainty. I don't think I realised quite how uncertain and anxious I was feeling, until now. But actually I don't know what I'm doing, and neither does 'society' when it comes to climate change.

> *Frankie - 'Not enough people are doing stuff about it, and they don't understand'*

CH reflection during interview—I can follow this and follow her narrative here with the next thing I say. I'm interested that she has started to use the word 'they' as if there was some group or 'other' who needed to be communicated with around climate change, but she has moved it from herself to 'the other' as soon as she starts to talk about the 'doing/action' aspect of climate change.

CH reflection after—When moving from her feelings to action she has distanced herself from her narrative and makes it someone else's problem. Separate from herself; I was curious who 'they' were. I was curious about the 'stuff' that they could be doing and could have asked her about this as well, but again I was slightly worried that she might try to come up with a specific answer and start to 'problem solve' climate change or look for solutions to it as a problem, and I wanted a more explorative and free association approach than this.

CH - What is it that they don't understand?

Frankie - 'The dangers of climate change and how the animals feel about it, like the Polar Bears'

CH reflection during interview—Oh, ok, she has gone straight to question 2 and 3 in her answer to Question 1, so I don't really need to ask these questions now. I lost the script already. On the one hand that's great, but on the other hand I'm feeling a bit lost. She has quickly linked climate change in her own mind to the impact on 'others' and also empathised with polar bears, she has an immediate 'relatedness' in her response to how the animals 'feel' about it. I don't think an adult participant would reflect on how animals 'feel' about it immediately like this unless asked directly, they may say what they feel as a person, or talk about the impact of climate change on animals, but not what the animals feel about it. She has started to use how polar bears might feel about climate change as a way of her talking about how she feels about climate change.

CH reflection after—This reminds me of the climate change 'Radical Hope & Cultural Tragedy' Climate Psychology Alliance Conference (18 April 2016) presentation by Jay Griffiths when I heard that we needed to use the intelligence of nature to guide the way when your world is under siege. That First Nation Americans used to include animals in decision-making asking 'who speaks for the wolf?' when making decisions. Kindness is related to the idea of kin, and that we are the kin of animals. We need an ethic of kindness in our laws and decisions and to align the human self with the law of nature.

I decided to stay with the image of the polar bear, to follow the image, to allow the image to start to speak for itself.

CH – Can you tell me why you chose a polar bear to talk about?

Frankie – 'Can't remember why, they're cute, and don't have enough ice so their numbers will go down'.

CH reflection during interview—It's interesting that her answer gives two things, both her feelings about polar bears where she sounds like a young child; but also a comment about climate change and the impact on polar bears due to the reduction in ice because of global warming where she sounds much more adult.

CH reflection after—The first half of her answer reflects on her feelings, but then she shifts to care and concern for others.

I decide to ask her the question about personification about climate change rather than continue on the polar bear theme, I don't want her to start to think that she has to be knowledgeable about polar bears.

CH reflection after—I was also interested to see if moving towards metaphor rather than the literal polar bears she has been talking about would evoke a different sort of response from her and take us in a different direction towards more unconscious communication. I think the polar bear narrative is more conscious and I was curious about expression of what was unconscious.

CH – If climate change were an animal, bird or plant, what sort would it be?

Frankie - 'Maybe a kind of bug or something that infests, like a fly or mosquito'.

CH reflection during interview—She answered this question really quickly; she hasn't hesitated or stopped and thought at all before answering this question. I'm also quite 'thrown' by her answer; it's not what I expected. I'm excited and slightly in awe of her imagination. I need to stay with the image of the bugs.

CH reflection after—The speed of her answer suggested that this wasn't an answer taken from conscious thought. I had expected more fluffy answers (such as more polar bears). Her use of bugs to personify climate change was a sophisticated and exciting metaphor. I had a sense of 'something' coming into the space that neither she nor I consciously decided on. I felt a perceptible shift, it reminded me of my practice as a psychotherapist when it was almost as if there was a subtle change in the room when the client shifted into less guarded, less defended and less conscious speech, the language of metaphor. I'm aware that I felt scared at the time wishing not to get back to more expected images or 'get back on track' with planned questions. I think we had entered a different relational field. I had never come across bugs as a metaphor for climate change before, I wasn't sure at the time if this personification of climate change as bugs would be unique to this child, or be reflected in further interviews.

CH - That's interesting, can you tell me why you chose that?

Frankie – 'Flies annoy people and mosquitoes bite people and that gives them energy, so like climate change takes energy to make it bigger'

CH reflection during interview—I'm intrigued and want to ask her to expand and say more about her choice of bugs to personify climate change.

CH reflection after—She had not tried to give me a rational answer or to justify or explain her answer; she stayed with her image and clearly had her own understanding about why she chose this as an image or personification of climate change. She was confident about her bugs as personification of climate change.

Frankie – 'Why did I say that? Because people generally don't like them, they're not nice and cause bad things to happen'

CH reflection during interview—I really want to know now what they would say if asked and how do they cause bad things to happen?

CH reflection after—I notice that we had the use of 'they' again in the narrative, but this time in relation to the bugs. I also noticed that the bugs and mosquitos were labelled as 'not nice'. So they had been chosen to represent a negative. In contrast, I wonder if polar bears are 'nice'? But I wanted to see if they have a voice and if she could speak through them or give them their voice?

CH – 'What do you think they would say about climate change?'

Frankie – 'Maybe ...... you've made the world a worse place so I'm going to do that too, as revenge'

CH reflection during interview—this is all a much darker/shadowy response than I expected.

CH reflection after—When the bugs as personification of climate change are given a voice, they speak of destruction and revenge, emotions that people often avoid, deny or edit from their narratives as they are frightening, shameful, raw, antisocial, wild and untamed. I think the use of personification has allowed this child to voice some very shadowy things, without shame or hesitation about how this might be heard or judged. In my mind this perfectly opened up a space where we could talk about climate change in terms of catastrophe, destruction, annihilation, extinction and revenge of the earth towards humankind; taboo and frightening narratives.

Later Reflections

I noticed that I 'squashed' the bugs' when I lost the free association around the image later in the interview, maybe this reflected a collective or archetypal horror of these creatures, a lack of respect for the 'small' and maybe also reflecting my own fears about climate change (and bugs maybe). I was reminded that we should 'make a place for the soul of the work to speak beyond the calculus of a researcher's subjective prejudices' (Romanyshyn 2010, p. 44).

In the weeks following this interview UK newspapers reported a huge fall in flying insects described as an 'Insect Armageddon', *The Guardian* (19 October 2017) talked about flying insects declining by 76% in 27 years. I started to reread Lauck (1998) 'If we learn to respond compassionately to insects, we will have deepened our capacity to respond compassionately to all species, including our fellow human beings. We will also have developed a discriminating eye towards the kind of propaganda that feeds our fears, and we'll have understood that we have the capacity to reinvent ourselves as a compassionate, cooperating species' (Lauck 1998, p. xvii). I also became much more aware how I dealt with bugs in my home, should I squash them? Or try to learn to listen to their message? I had to think about my ambivalence towards mosquitos, wasps and hornets; and I struggled with a heightened awareness of my own need for control and to rid my house of horseflies.

Examples and Reflection on Themes from Other Research Interviews

I was interested to see if this methodology would evoke similar themes from interviews with other children; I will briefly introduce two key emerging themes; empathetic relationship with (non-human) nature and climate change personified as a destructive vengeful force.

Most children replied unprompted to a general opening question about climate change not by commenting on the impact on people or themselves as children, but by talking about the impact on animals or the natural (non-human) world: *'it's not good, animals are dying', 'the coral is bleaching and dying', 'the fish are missing', 'I'm sad, my friends the fish are dying'.* The distress caused in particular to children by loss of connection with the natural (non-human) world has been written about extensively by (to name a few) Abram (1996), Louv (2005), Monbiot (2017), Totton (2011), Macfarlane (2015), and Griffiths (2013), and although I was using relationship with nature to talk about climate change in this research, so could be guilty of evoking some of these responses consciously or unconsciously, what was interesting to me was that these children seemed to be spontaneously thinking about climate change in relation to animals and nature. My analysis of this is that although there may be considerable disconnection, confusion and loss of relationship to nature in everyday life for many, the children I interviewed seemed to be alive to their relationship with nature imaginally as well as empathetically.

The personification of climate change as an animal, bird or plant produced varying images, but all with narratives of destruction; a lion said *'I'm sorry I've done this, I'm killing other animals and they're extinct'*, a very large lizard *'I'm hungry, and my appetite will never be satisfied'*, a shark (said sharkily) *'I don't want to eat you but there's not much else left to eat now, I'm coming now'*, a Tazmanian Devil said *'we will steal food from anyone to survive this'*; and from the plant world, brambles and creepers *'we wrap around a plant and kill it and flourish ourselves because the plant is dead'*.

Some of the children's responses were notably different to my expectations; particularly personification of climate change as a small creature (bug) contrasting with the images I had in mind at the start of this research; others seemed more consciously close to popular narratives about climate change, such as sharks and polar bears. In all cases, it seemed to me that the children were spontaneous and unrestrained in how they personified climate change. They were imaginative and had no problem in giving voice to destructive thoughts and attitudes through personification.

When asked at the end of the interview if they had a message for adults about climate change, I repeatedly heard clear anger and despair: *'adults think we are stupid, but we see what's going on really'; 'nature is fighting back, and you don't see it, we do'; 'it's the ignorance of adults who don't listen to the children'; 'people have lost sight of what's really important'; 'stop ruining young people's lives'.*

Returning to the argument of Currie and Deschênes (2016) that children should be central to climate change debates, this methodology seems to be effective in helping children talk about how they feel about climate change Perhaps it could also help children voice some unbearable truths for adults about the destructive human impulses that have led us to be in this position of human-induced climate change in the first place, and remind us to listen to the insects, fish, plants and animals, as well as to the children.

Final Thoughts

Barack Obama on June 25, 2013, making his Climate Action Plan, argued that our children in the future would look back at us and ask if we did all that we could have done to deal with this problem, or did we avoid doing what needed to be done? With climate change sceptics gaining voice I would argue that this research into children's views might unfortunately be needed more now than ever before. Maybe we also need to listen more carefully to the sharks and the creepers, the insects, the whirr of small wings, the rustle of many feet, the bugs as well as the 'aaarrrggghhhhs'.

Why am I doing this?
Only the tiny sea of my cells replies,
reminding me that
I am the sea and the sea is in me.

(Jacques-Yves Cousteau, 1953)

References

Abram, D. (1996). *The spell of the sensuous*. New York: Vintage Books.

Alvesson, M., & Skoldberg, K. (2000). *Reflexive methodology: New vistas for qualitative research*. London: Sage.

Bering, J. M. (2006). The folk psychology of souls. *Behaviour and Brain Science, 29*, 453–462.

Clarke, S., & Hoggett, P. (2009). *Researching beneath the surface: Psycho-Social research methods in practice*. London: Karnac Books.

Clayton, S., Leubke, J., Saunders, C., Matiasek, J., & Grajal, A. (2014). Connecting to nature at the zoo: Implications for responding to climate change. *Environmental Education Research, 20*(4), 460–475.

Cooper, A. (2009). Hearing the grass grow: Emotional & epistemological challenges of practice near research. *Journal of Social Work Practice, 23*(4), 429–442.

Cousteau, J. (1953). *The silent world*. London: Hamish Hamilton Ltd.

Currie, J., & Deschênes, O. (2016). Children and climate change: Introducing the issue. *The Future of Children, 26*(1), 3–9.

Darbyshire, P., Schiller, W., & MacDougall, C. (2005). Extending new paradigm childhood research: Meeting the challenges of including younger children. *Early Childhood Development and Care, 175*(6), 467–472.

Frantz, C., Mayer, S. F., Norton, C., & Rock, M. (2005). There is no 'I' in nature: The influence of self-awareness on connectedness to nature. *Journal of Environmental Psychology, 25*, 427–436.

Gibbons, E. D. (2014). Climate change, children's rights, and the pursuit of intergenerational climate justice. *Health & Human Rights, 16*(1), 19–31.

Griffiths, J. (2013). *Kith: The riddle of the childscape*. London: Penguin Books.

Hamilton, C. (2013). What history can teach us about climate change denial. In S. Weintrobe (Ed.), *Engaging with climate change: Psychoanalytic and interdisciplinary perspectives*. London: Routledge.

Hillman, J. (1975). *Re-visioning psychology*. HarperCollins.

Hollway, W., & Jefferson, T. (2013). *Doing qualitative research differently: A psychosocial approach*. London: Sage.

Holmes, J. (2013). A comparison of clinical psychoanalysis and research interviews. *Journal of Human Relations, 66*(9), 1183–1199.

Jung, C. G. (1960). *On the nature of the psyche, collected works, Vol. 8*. New York: Bollingen Foundation.

Kwan, V. S. Y., & Fiske, S. T. (2008). Missing links in social cognition: The continuum from nonhuman agents to dehumanized humans. *Social Cognition, 26*, 125–128.

Lauck, J. E. (1998). *The voice of the infinite in the small: Revisioning the insect— Human connection*. Mill Spring, NC: Swan Raven Publishing.

Lertzman, R. (2015). *Environmental melancholia: Psychoanalytic dimensions of engagement*. London: Routledge.

Louv, R. (2005). *Last child in the woods: Saving our children from nature deficit disorder*. Chapel Hill, NC: Algonquin Books.

Macfarlane, R. (2015). *Landmarks*. London: Penguin Books.

Monbiot, G. (2017). *Out of the wreckage: A new politics for an age of crisis*. London: Verso.

Ojala, M. (2012). Regulating worry, promoting hope: How do children, adolescents, and young adults cope with climate change? *International Journal of Environmental & Science Education, 7*(4), 537–561.

Phelan, S. K., & Kinsella, E. A. (2013). Picture this….safety, dignity and voice—Ethical research with children: Practical considerations for the reflexive researcher. *Qualitative Inquiry, 19*(2), 81–90.

Romanyshyn, R. D. (2010). The wounded researcher: Making a place for unconscious dynamics in the research process. *The Humanistic Psychologist, 38*(4), 275–304.

Romanyshyn, R. D. (2013). Making a place for unconscious factors in research. *International Journal of Multiple Research Approaches, 7*(3), 314–329.

Schaverien, J. (2005). Art, dreams and active imagination: A post-Jungian approach to transference and the image. *Journal of Analytical Psychology, 50*, 127–153.

Stanley, F., & Farrant, B. (2015). Climate change and children's health: A commentary. *Children, 2*, 412–423.

Stewart, W. (1998). *Dictionary of images and symbols in counselling.* London and Philadelphia: JKP.

Strømme, H., Gullestad, E., Stanicke, E., & Killingmo, B. (2010). A widening scope on therapist development: Designing a research interview informed by psychoanalysis. *Qualitative Research in Psychology, 7*(3), 214–232.

Tisdall, E. K. M., Davis, J. M., & Gallagher, M. (2009). *Researching with children & young people: Research design, methods and analysis.* London: Sage.

Totton, N. (2011). *Wild therapy: Undomesticating inner & outer worlds.* Monmouth: PCCS Books Ltd.

Weintrobe, S. (Ed.). (2013). *Engaging with climate change: Psychoanalytic and interdisciplinary perspectives.* London: Routledge.

White, H. (1980). The value of narrative in the representation of reality. *Critical Inquiry, 7*, 5–27.

Xu, Z., Sheffield, P. E., Hu, W., Su, H., Yu, W., Qi, X., & Tong, S. (2012). Climate change and children's health—A call for research on what works to protect children. *International Journal of Environmental Research & Public Health, 9*, 3298–3316.

4

An Integrative Methodology for Investigating Lived Experience and the Psychosocial Factors Influencing Environmental Cognition and Behaviour

Nadine Andrews

Introduction

A great deal of research and policy attention has been devoted to changing the behaviour of individuals in order to advance climate change mitigation and protection of species and habitats. However, I am interested not in behaviour change or even behaviour itself, but its underlying drivers specifically the psychosocial factors influencing individuals who are already highly motivated to behave in pro-environmental ways but don't—or at least not all the time or in all aspects of their lives. For if people with strong pro-environmental values do not enact these values consistently, what hope for changing the behaviours of people with weak commitment to these values? In this chapter I describe the approach I took to research this area, the so-called values-action gap, focussing on a study that enquired into the lived experience of

N. Andrews (✉)
The Pentland Centre for Sustainability in Business, Lancaster University, Lancaster, UK

© The Author(s) 2019
P. Hoggett (ed.), *Climate Psychology*, Studies in the Psychosocial,
https://doi.org/10.1007/978-3-030-11741-2_4

individuals working to influence and improve environmental decision-making in their organisations.

Various factors have been proposed to explain incongruence between values and behaviour but the models produced are fairly limited. This is because the studies tend to be decontextualised and deal with different factors[1] in an isolated rather than integrated way (Clayton et al. 2016; Lülfs and Hahn 2014). Many studies rely on theories such as the Theory of Planned Behaviour that are ill-equipped to engage with the complexity of lived experience, particularly the non-rational, unconscious and context-specific dimensions (Kollmuss and Agyeman 2002). Rather than avoiding complexity, my methodology was designed to embrace it. Situating the research participants in the context of their work settings and the dominant cultural worldview of the society in which they and their organisations are embedded, I investigated psychosocial factors influencing their cognition and actions, drawing on concepts and ideas from various fields including systems thinking, social and ecopsychology, cognitive and ecolinguistics, environmental philosophy and organisational studies. This transdisciplinary methodology enabled me to identify and model the dynamics of multiple cross-level factors and produce nuanced in-depth findings. This would not have been possible with a typical survey-based approach (see Andrews 2017a, b). The factors I identified all relate directly or indirectly to psychological threat coping strategies:

- Sources of tension and threat
- Coping strategies used to negotiate these tensions
- Outcomes of coping strategies (ecologically adaptive or maladaptive implications)
- Various factors influencing the efficacy of coping strategies
- Contextual factors (organisational and socio-cultural) influencing one or other of these aspects.

---

[1]Factors are typically called 'determinants' in this literature, illustrating a linear causal mode of thinking about relationships.

Following an explanation of the philosophical approach underpinning the study, I describe the procedure I adopted to generate, analyse and interpret data. I then provide a worked example of how findings from this process can be written up, generating insights into the phenomenon under study. Finally, I reflect on the implications of the findings for climate and environmental action. Together, these aspects offer a rich and detailed account of the methodology and demonstrate its value as an approach for engaging with the complexity of lived experience and investigating underlying drivers of environmental behaviour.

Philosophical Approach

The methodology I designed proceeds from four main premises and assumptions relating to: embodied and situated experience, unconscious processes, language as a window into conceptual systems, and my own ecological philosophy.

First, the methodology rejects the dichotomy that is often made between objectivism and subjectivism, and between our outer and inner worlds. To form this understanding I brought together three epistemological approaches: Interpretative Phenomenological Analysis (IPA), psychosocial studies, and the embodied realism of cognitive linguistics.

IPA is concerned with gaining insight into how someone experiences and makes sense of a given phenomenon in a particular context through intersubjective inquiry and analysis. IPA recognises that we cannot remove our conceptual systems from our observations of the world to find out how things "really are" in an objective sense (Larkin et al. 2006). IPA sees individuals as essentially embedded, intertwined and immersed in the world of objects and relationships, language and culture, projects and concerns (Larkin et al. 2006; Smith et al. 2009). Experience is a lived process, an "unfurling of perspectives and meanings, which are unique to the person's embodied and situated relationship to the world" (Smith et al. 2009, p. 21). In IPA research, there is, therefore, a commitment to situating personal meaning in socio-cultural contexts (Eatough and Smith 2010). More than that, it is considered a criterion for assessing quality and validity (Smith et al. 2009).

This epistemological position is compatible with the epistemology of embodied realism proposed by cognitive linguists and philosophers Lakoff and Johnson. Truth, they propose, is always relative to a conceptual system, grounded in and continually tested by our ongoing embodied experiences of interacting with the world (Johnson and Lakoff 2002; Lakoff and Johnson 1999). For example, "a clearing is not an inherent property of that place in the woods where the trees are less dense but a property that we project onto it relative to the way we function with respect to it" (Lakoff and Johnson 1980, p. 158). According to Johnson (1987), one of the pervasive features of human experience is the experience of physical containment, which I discuss later in the example of analytic commentary.

The situated aspect of experience is fundamental to the perspective of psychosocial studies, which holds that for the individual, the psychological and the social are always intertwined and cannot be meaningfully separated (Woodward 2015). Indeed, "the social unconscious operates according to principles that are similar to those of the individual unconscious" (Woodward 2015, p. 142). A focus on the unconscious is a central feature of psychosocial studies (Clarke and Hoggett 2009).

IPA, cognitive linguistics and psychosocial studies all recognise human experience as embodied and situated, involving fundamental entanglements of psychological processes and social contextual forces. In both cognitive science and phenomenology, the physical body is regarded as integral to perception and understanding experience: it is not just a biological object but a lived experiential structure (Lakoff and Johnson 1980; Varela et al. 1993). This understanding about how humans create meaning forms the foundation upon which my analysis and interpretation is built. In my study, the research subjects are situated in their work settings (person-in-organisation) and in the macro-context of the dominant cultural worldview of the Western industrialised societies in which they, and their organisations, are embedded (person-in-society). The meanings they make emerges out of their ongoing embodied interactions with their work colleagues, organisation and society.

The second feature of the philosophical framework is the premise that the mind has conceptual systems that structure perception, understanding and behaviour, that this occurs in mostly unconscious processes, and that people may not, therefore, be consciously aware of all the processes involved in their behaviour and experience (Lakoff and Johnson 1980; Willig and Stainton-Rogers 2010). This is why descriptive analyses of self-report surveys have limited explanatory power. IPA permits the researcher to develop an alternative narrative of the research subject's experience, informed by extant theory. I drew upon Self-determination Theory (SDT; Deci and Ryan 2000), particularly constructs of basic psychological needs, intrinsic and extrinsic goals, controlled and autonomous motivation, and vitality—the energy available to the self for action. These were themes that emerged as I analysed the data, and I found SDT useful in explaining these results. In the example of a narrative I provide later, I discuss the findings in relation to competency, autonomy and relatedness needs. Competency refers to feeling competent and effective, engaging in optimal challenges, and having an effect in the world as well as attaining valued outcomes within it. Autonomy refers to the desire to self-organise experience and behaviour, and for activity to be in accord with one's integrated sense of self and intrinsic choices. Relatedness involves seeking attachments and experiencing feelings of security, belongingness and intimacy with others. SDT proposes that these needs are drivers of behaviour and require satisfying for psychological wellbeing (Deci and Ryan 2000).

The third main premise of the methodology is that our conceptual systems are fundamentally metaphorical in nature, and that these metaphors are expressed in language. Metaphor use is therefore not arbitrary—rather, by a person's use of particular terms, something can be inferred about how they consciously or unconsciously conceptualise their world (Deignan 2005; Lakoff and Johnson 1980). Language both reflects and shapes the way we think, influencing how we act and relate to the world. We are continually being primed to conceptualise the world in particular ways by the dominant discourse of our social contexts, and it is this feature that makes language psychosocial (Andrews 2018). In my study, I analysed the participants' accounts for frames and metaphors. This is a form of micro-discourse analysis, which means coding the text line-by-line, often word-by-word. Integrating IPA with frame and metaphor analysis enabled me to explore below the surface of the research subjects' conscious awareness.

The final element I wish to make explicit is my own ecological philosophy, which informs how I analyse and interpret data and draw conclusions about the implications of the findings for climate and environmental action. Following others (e.g. Kidner 2001; Merchant 1983; Plumwood 1993; White 1967), I believe that perceiving humans as separate from and superior to nature is a root cause of ecological crisis. This refers not just to the external world of plants and animals and natural forces but also to our inner world and the aspects of the self often associated with nature and wildness: emotions, the physical body, intuition and the unconscious mind (Hasbach 2012; Rust 2008; Totton 2011). There are two key implications of this philosophy:

- The "social" in psychosocial has to include relationship with other species, for the world in which we are embedded includes the other-than human natural world.
- Restoring healthy ecological balance of the planet requires restoring a healthy relationship between parts of the self and dissolving dualistic relationships between mind and body, reason and emotion.

Methodology for Conducting Fieldwork

Recruitment and Selection of Participants

In IPA, research subjects are selected on the basis that they grant access to a particular perspective on the phenomena under study: this is known as purposive homogenous sampling. In my study the phenomenon was the experience of an individual oriented to pro-environmental values working in a formal role to influence organisational practices, and the socio-cultural context for this phenomenon is industrial growth societies where nature is predominantly framed in economic terms (Andrews 2018).

I recruited five sustainability managers and a chief executive to the study. The participants had formal roles ranging from production of environmental strategies and policies, delivery of energy efficiency programmes and conservation of habitats. The organisations were in local and regional government, social housing, credit union and health care sectors, in the UK and Canada. The participants were recruited via a variety of channels including professional networks and social media, and were selected according to three criteria to satisfy the requirement for homogeneity:

- Currently in employment with a formal role to influence environmental decision-making
- Strong pro-environmental values and goals motivating them to do their work
- Propensity and willingness to be mindful (so that they would be able to give rich accounts of their subjective experience).

IPA studies tend to focus on a small number of people (Larkin et al. 2006). This is because IPA recognises the value of the single case and has an idiographic focus which means detailed examination of the research participant as a particular unique individual. This requires working with data in a highly detailed and intense way, which generates a lot of material and is very time-consuming. A sample of six would render the findings invalid if it were quantitative research, and would also be considered too small in some other qualitative approaches such as grounded theory. However, in IPA the small sample does *not* mean that nothing meaningful can be concluded about the phenomenon—on the contrary, something very tangible and real about the subject matter can be revealed (Larkin et al. 2006).

Data Collection

The main method for collecting data in IPA is semi-structured interviews, with diaries and focus groups also used (Smith et al. 2009). The methods used in my study are shown in Table 4.1, together with their rationale. Interviews were my main data source, and I focus on explaining the procedure for analysing this data next.

Table 4.1 Data sources

Data sources	Format	Rationale
2-hour in-depth semi-structured interviews	Audio recorded and transcribed verbatim	Obtain detailed account of experience, can probe deeper and follow emergent lines of enquiry
Participant diary of experience of significant meeting	Entries written shortly after important meeting by five participants	Record embodied awareness of the experience
Researcher reflexive diary	Entries written shortly after each encounter and ad hoc throughout	Record my experience and reflections to ensure rigour and quality
Indirect observation of participant	Audio recordings by two participants of a meeting with colleagues	Indication of organisational discourse and salient frames that participants are exposed to or are interacting with
Organisational documents relating to environmental policy, strategy or practice	Public online resources and documents, and internal documents e.g. meeting minutes, briefing notes	
Final debrief conversation	Audio recorded	Share theoretical framework and key findings with participants as a stakeholder check for credibility and trustworthiness of findings (Thomas 2006)

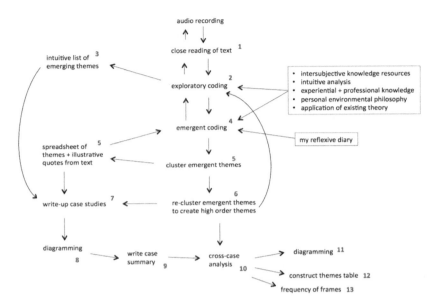

Fig. 4.1 Analysis procedure

Data Analysis and Interpretation

IPA offers a flexible framework for analysing data. These stages are not remarkable compared to other qualitative methods: the difference is in the philosophical approach as described above and in allowing intuitive, experiential and professional knowledge to be drawn upon (Smith et al. 2009). I adapted the framework, developing a nonlinear procedure involving numerous iterative cycles as shown in Fig. 4.1.

Each interview was transcribed verbatim including notes of nonverbal elements such as laughter, sighs and pauses.

Stage 1—close reading of text:
Pseudonyms are assigned to each participant. Verbatim transcription of interview, number each line, several close readings of the text.

Stage 2—exploratory coding:
This is an open process that involves asking descriptive, linguistic and conceptual questions of the text (Smith et al. 2009), as shown in Table 4.2.

Table 4.2 Exploratory coding in IPA

Descriptive Qs
- How can this sentence/phrase/idea be summarised?
- What experiences are being described and claimed by the participant?
- What are the key features of those experiences for the participant?

Linguistic Qs
- How is the participant saying what they are saying (hesitant, forceful, stumbling)?
- Is the participant shifting tenses or pronouns in their speech, what might this mean?
- Has the participant used any metaphors or interesting phrases?

Conceptual Qs
- What is this person trying to say?
- What is going on here?
- What is underlying this?
- What do the experiences appear to mean for the participant?
- So what (what = the impact of what is said)?
- What does X mean, what does Y mean?
- What is missing or not being said?

As we can see, metaphors are classed as linguistic but in cognitive linguistics metaphors are not primarily linguistic but cognitive: they activate structures in the brain (frame-circuits) that serve as frames of reference for interpreting new information and experiences. Nevertheless, it is here that frame and metaphor analysis can be integrated into IPA to form a coherent methodology.

When we use metaphors we are thinking about one thing (the target domain) in terms of another (the source domain), and knowledge about the source domain is used in reasoning about the target domain—this knowledge is the entailment of the metaphor. For example, let's consider the phrase 'spend time'. Here, time is being conceptualised in terms of something that is 'spent'. Typically we associate spending with money: so what we know about money (the source domain) is mapped over onto how we think about time (the target domain). As money is a resource of financial value we might, therefore, think of time as a resource of financial value. 'Spend' is the trigger word that activates a financial transaction frame-circuit in the mind. It is these trigger words that the analyst looks out for in a text. However, as the use of frames and metaphors is often unconscious (Lakoff and Johnson 1980), this means trigger words are not always immediately obvious so metaphors can be easily overlooked in a text. Deignan (2005) emphasises the importance of establishing consistent procedures for spotting metaphors. There is no single correct way to conduct frame and metaphor analysis. I used various established knowledge resources to guide identification of frames and metaphors and their likely entailments. Key resources included dictionaries (e.g. OED Online and Etymonline) and frame and metaphor databases (e.g. FrameNet, Metalude).

I coded the interview transcripts for various aspects of experience as shown in Table 4.3 that the literature I reviewed indicated could be influencing factors. However, I did not know *whether* these would, in fact, show up nor *how* they would show up.

Metaphors are incomplete representations of reality that privilege one way of seeing and obscure other ways. Using abductive reasoning, the analyst asks *what is being promoted and what is being hidden, and what are the implications?* It is here, explicitly or implicitly, that the analyst's personal ecological philosophy has a significant influence on the conclusions that are drawn.

Stage 3—intuitive identification of key themes:
This involved listing whatever struck me as the main "objects of concern" (Larkin et al. 2006) for the participants. The intuitive insights came from noticing signals such as body language, tone of voice, hesitations, laughter and so on. I referred to these notes later in Stage 6 and checked them against the results of the systematic analysis.

Stage 4—emergent coding:
I then used a more systematic, controlled and deliberate form of analysis to identify salient themes from my exploratory coding notes for each page of transcript and listed them separately. Once this was complete for each case I revisited these lists and looked for patterns, and then compiled a refined version of key themes for each case.

Stage 5—cluster emergent themes:
This stage involved systematically going through the exploratory notes and emergent coding notes in a second cycle of analysis and re-clustering of emerging themes.

Table 4.3 Analysis focus

Analysis focus	Rationale
Frames and metaphors	To gain insight into (a) how participants conceptualise their experience, as this has influence on behaviour (e.g. Lakoff 2010), and (b) how the organisation conceptualises the natural world
Values, goals and identity salience	Which type of values, goals and identities that are salient has implications for the kind of behaviours that are motivated (e.g. Crompton and Kasser 2009)
Defences/coping strategies	Strategies used to cope with psychological threat may be ecologically adaptive or maladaptive (Crompton and Kasser 2009; Weintrobe 2013)
Emotion, emotional management	What people feel and how they accept, express and regulate those emotions has implications for behaviour and for well-being (Weintrobe 2013; Rogelberg 2006). Emotion is central to understanding human experience (Smith et al. 2009)
Mindfulness and embodied cognition	Cognition is embodied and situated (Lakoff and Johnson 1980; Varela et al. 1993). Mindful awareness supports autonomous self-regulation and helps resist contextual pressures (Deci et al. 2015)
Needs satisfaction, type of motivation, and vitality	The extent and manner in which basic needs are satisfied, and type of motivation, have implications for vitality and hence effectiveness (Deci and Ryan 2000; Ryan and Deci 2008). Type of motivation also has implications for the likelihood of a behaviour being performed, as well as how well it is performed
Conflicts and ambivalence	How inner conflicts are negotiated has implications for effectiveness (Wright et al. 2012). Inner coherence and identity commitment support values-congruent behaviour (Deci and Ryan 2000). Ambivalence can impede capacities for concern and repair (Lertzman 2015), but may also be tapped as source of strength and vitality (Meyerson and Scully 1995)

Stage 6—re-cluster emergent themes into higher order themes:

I created higher order themes by re-clustering emerging themes derived from the systematic analysis of Stages 4 and 5 and compared them with key themes derived from intuitive analysis in Stage 3. This clustering process in this stage was guided by the focus of the research questions, which meant I could narrow down the themes to those that were key in the participants' experience of influencing pro-environmental decision-making in the organisation. Through this process I arrived at seven higher order themes, each with a set of associated subthemes (see Andrews 2017b).

Stage 7—write up case study:
I wrote a case study for each participant using content from spreadsheet, emergent coding notes, my reflexive diary and intuitive analysis notes. The format was summary of role in organisation followed by a narrative by higher order theme that described how each theme manifested for the participant.

Stage 8—diagram single cases:
I sketched diagrams of key features of participants' situations, as I interpreted it. Diagrams are a systems thinking tool for abstracting and representing aspects of complex situations in simplified form and showing the interconnectedness between different parts. Diagramming as a process helped me with sense making and gaining clarification about feedback loops in the situation. The diagram as an object is a visual representation of findings that can also be used in communicating the findings.

Stage 9—write case summaries:
Summaries were then created from the case studies. The format for this was a table with the themes in rows and the participant information in columns, which enabled me to easily see similarities and differences between the cases.

Stage 10—cross-case analysis:
I then performed cross-case analysis using information from the case studies and the case summary table. This involved determining where there were similarities and differences in the way that higher order themes manifested for each case, and identifying areas of convergence and divergence in exploratory themes. I checked back to the clustered themes (Stage 5) for the entries for each participant to make sure nothing significant was missed. There was some refining of the phrasing of

higher order and exploratory themes as a consequence of the cross-case analysis.

Stage 11—diagram across cases:
Using an inductive process, I sketched diagrams modelling relationships between key themes based on synthesis of interpretations across cases, with reference to the diagrams produced in Stage 8 (see Andrews 2017b).

Stage 12—construct themes table:
I then created a table of key higher order themes and associated sub themes (emerging themes) with illustrative quotes (see Andrews 2017b).

Stage 13—frequency of frames:
I counted the frequency of particular trigger words as a way of assessing salience of associated frames. This was only intended to be indicative, as I did not capture all the possible words or phrases that could be associated with a particular frame. Frequency of use may suggest salience but it doesn't necessarily—tone of voice and body language are also cues.

Constructing a Narrative

Once the stages of analysis are complete, the final step is to write up the findings, interweaving extracts from participants' accounts with analytic commentary informed by the literature. The extracts are the phenomenological evidence grounding the interpretation (Smith et al. 2009). Typically, the findings are presented by higher order theme.

To demonstrate how a written narrative can be constructed, I will focus on the higher order theme 'Engagement with negative emotion' with subthemes 'emotion regulation' and 'reason-emotion dualism', all of which emerged in my research. Emotion is an embodied experience—it is a physiological and mental feeling state, involving physical sensations of arousal and cognitive appraisal of the cause of arousal in response to stimuli. Emotion regulation is the ability to control and productively use one's emotions, including the processes by which individuals influence which emotions they have, when they have them and how they experience and express them. Reason-emotion dualism refers to a perceived hierarchical split between the rational mind and the emotional body, an idea that took hold in the Scientific Revolution and remains dominant still (Merchant 1983; Midgley 2003). Emotions are an important focus of study in climate psychology because of the powerful role that they play in directing attention and guiding behaviour (Stangor 2010).

The narrative I present is illustrated with extracts from the interview with one of the participants, Ash, a sustainability manager at a local authority in the UK whose work involved delivering renewable energy programmes. Quotes are written in italics.

Emotion Regulation

I ask Ash how he feels about the environmental situation:

> Ash: *Emotions I suppose when I think about it which I try and avoid think-it's rare that I- it makes me feel sad. Uh (pause)*

> Me: *Mostly sad?*
> Ash: *Frustrated. Not really angry, not really angry. I think I probably would have done 20 years ago. I think I just feel a sense of melancholy about it really (pause)*
> Me: *So what do you do with those emotions?*
> Ash: *Um (pause) I tend not to explore them or I think I've got them in a box in my head and I carry on um being a parent and being a sustainability manager and trying to do good stuff. So and I kind of I you know in a box sitting in an attic in my mind really so I don't think its particularly helpful to explore because it's kind of disabling in a way really and disheartening. So I think I put it in a box in the attic*
> Me: *Does it still have an effect on you?*
> Ash: *From time to time when I think about it yeah. Yeah. But it's literally in the back of my mind I suppose. So maybe it does tinge. No I'm not defeatist in my job so but yeah a sense of melancholy I suppose*

In this extract, Ash mentions several emotional states: sadness, frustration, "not really angry", melancholy, disheartened, "not defeatist". But it seems that these emotions are being suppressed and avoided: there are descriptive self-reports of avoidance and compartmentalisation (e.g. *box in the attic*). This requires effort, as indicated by *try*. Ash talks about what he thinks rather than how he feels—a type of intellectualisation that serves to distance the speaker from the felt emotion, and there are linguistic cues of pauses and not finishing sentences. Climate change and ecological crisis pose profound psychological threat (Crompton and Kasser 2009), so these forms of emotion regulation can be interpreted as ways of defending against this threat. It would appear that for Ash, avoidant coping is both a conscious and unconscious strategy.

I will return to discuss the *box in attic* from a cognitive linguistics perspective, but first I draw on self-determination theory to gain further insight into the threat that is triggering Ash's avoidant coping response. Interpreting the extract above through this lens reveals that there is an association in his mind between emotions of sadness and melancholy with competency, and that these emotions pose a threat to the satisfaction of his competency needs. This is indicated in several different ways:

Firstly, Ash describes feeling *frustrated*. Frustration is an indication of thwarted competency needs (Pryzbylski et al. 2014).

Secondly, *disabling*. This term indicates a belief that negative emotions have a disruptive effect on effectiveness, threatening satisfaction of competency needs. This supports Macy's (1993) assertion that "we are afraid that we might break apart or get stuck in despair if we open our eyes to the dangers" (p. 31).

Thirdly, *disheartening*. This term refers to a part of the body—the heart, often associated with emotion, spirit, desire and courage. Its use invokes not just a physical feeling in the body but also a weakening of resolve to act, a theme that recurs elsewhere in the interview (*give up*):

> *So certainly in a global context generally put myself in the pessimist camp* [...] *Generally I'm an optimist though, I need that otherwise I'd give up I think, so.*

Giving up is clearly an issue for the satisfaction of competency needs.

The fourth indication is the phrase *not defeatist in my job*. Defeat is associated with failure of action, and hence is a threat to competency. The term defeatist means a person who expects or is too ready to accept failure. As shown in the quote above, Ash uses optimism as a way to keep going. Indeed, being optimistic is part of his motivational story (*I need that otherwise I'd give up*) but optimism, if unrealistic, is a form of self-deception and an ecologically maladaptive coping strategy because it allows the individual to deny the reality of the situation and avoid responding accordingly.

The motivational story is the story that Ash tells himself that provides a rationale for the work he does, justifies his experience and motivates him to keep doing what he is doing. As Ash himself explains, he is *trying to do good stuff*. The story is therefore about competency but it also serves autonomy needs because the creation and maintenance of narratives such as this creates personal meaning and reconciles inconsistencies in sense of identity (Vignoles 2011), enhancing inner coherence. However, we can see from other statements Ash makes that autonomy needs are not fully satisfied by the story. Elsewhere in the interview are indications of a sense of being up against a more powerful force: he refers to the loss of natural spaces as a *slow inevitable irreversible decline,* and describes working with politicians to help them *swim against that tide* (of development) *as much as they're able to.* This highlights the internal tensions and contradictions that are part of his lived experience.

The Safety of the Container

I now return to explore the *box in the attic* phrase from a cognitive linguistics perspective. The *box* into which Ash puts his emotions is an image schema of containment. This schema is often used as a conceptual domain for safety and security and implies protection of what is in the container from dangers outside, or protection of what is outside from the danger within (Johnson 1987). This is close to the understanding of "containment" in psychoanalytic terms (Bion 1962). The danger, as I have just established, is thwarted competency. Now that Ash has realised that he has compartmentalised his emotions about ecological crisis into a box in the attic of his mind, and is opening up this box by talking to me about these feelings, they are now potentially uncontained and out of control. I had the impression he was keen to put them back in the box as quickly as the interview allowed. This theme of safety is reflected in other statements Ash makes. Ash has explained that most of his organisation's services have some sort of environmental impact, and I ask how he feels emotionally about the impact that his organisation is having:

> Ash: *(silence) how do I feel about it (in quiet voice)? As local government officers it's all bashed out of us in our day job because what we feel about things is completely irrelevant, it's about what the business case is, and you know pragmatic*

Ash conceptualises his organisation as violently hostile to emotions about ecological crisis (*bashed out of us*). A reason–emotion dualism is evident: rationality (*business case, pragmatic*) is privileged and feelings are deemed irrelevant. Lertzman (2015, p. 33) writes, "The capacity to be disturbed is linked with the capacity to be curious, and both require certain levels of containment and safety to help tolerate such experiences". For Ash, his organisation is an unsafe container for expressing feelings of sadness, melancholy and pessimism, making containment of these emotions within himself (*box in the attic*) the safer option. SDT recognises that social contexts can be supportive or undermining of needs satisfaction: dissonance between felt and expressed emotions is likely to be affecting Ash's sense of inner coherence.

Emotion and Identity

In the extracts discussed above, experience of emotion is closely connected to a sense of self. Ash separates exploration of difficult emotions about the ecological crisis from his identities of *being a parent* and *being a sustainability manager*. He also identifies with being both *in the pessimism camp* and as *an optimist*, which reveals inner tensions. In the last extract, he links the identity of local government officer with lack of emotion, which he associates with pragmatism and rational choice economics (*the business case*). This conceptualisation is a recurring theme in the interview. For example, Ash explains how influencing senior executives is critical to him being successful in his role, and this requires him to be seen as credible. He continues:

So early days fresh out of university very clear I'm not going to wear sandals bring lentil sandwich to work and have a beard [...] So being seen as being credible and professional [...]

Ash has deliberately chosen to suppress an identity in order to be perceived in a favourable way by those he wishes to influence—his pursuit of relatedness is in service of competency. However, the strategy doesn't work with everyone:

despite me being here for many years the Treasurer still thinks I'm basically resident Friends of the Earth [...] I could, I could call myself Ash Corporate, you know and I live at number 1 Corporate Street Corporate Town and I love doing corporate and just talk about- and he still wouldn't believe I was anything other than Friends of the Earth in residence and he's just got that despite my best efforts not to you know wear a green shirt and uh open toed sandals or try and speak with too much passion or emotion about why we need to act, which is why I find it difficult - because I so consciously shut that out.

This suppressed identity, which he refers to elsewhere in the interview as "deep green", is associated here with passion and emotion. The effort involved in suppression is clearly articulated (*despite my best efforts, I so consciously shut that out*). But this passion for Ash can be counterproductive at times. He states:

sometimes particularly in meetings with senior managers I try and rein it in.

Here, emotion is conceptualised as a wild animal that needs to be brought under control. The metaphor implies a dualism between reason and emotion, and associates emotion with wild nature. This reining in of emotion is a form of containment, and it takes effort as implied with *try*.

Implications for Adaptive Responses to Ecological Crisis

Ash suppresses negative emotions about the ecological crisis because they are perceived to be a threat to competency. Contextual forces in the form of an organisational culture of reason-emotion dualism further discourage Ash from expressing his felt emotions. But suppression of strong emotion takes mental and physical effort, depleting vitality and impairing the ability to think (Rogelberg 2006) so there are consequences. Emotional avoidance may be appropriate in the short term for example when it is not within a person's control to change a situation at that moment (Stangor 2010) but over the longer term, it is associated with poorer health (Weinstein and Ryan 2011). More than this, emotional avoidance also has direct implications for how we respond to the ecological crisis. Sadness, anger, melancholy, etc. are sources of information. Ignoring these cues protects us from having to make radical changes or take significant climate and environmental action. As Rust (2008, p. 160) says, "when we block out our feelings we lose touch with the urgency of the crisis". In the analysis example I have presented, we can see that emotional avoidance is a coping strategy that is both an underlying driver of behaviour and a factor influencing the relationship between pro-environmental values and action.

Conclusion

In this chapter, I have described a methodology for investigating conscious and unconscious processes in lived experience. The study for which the methodology was designed aimed to gain insight into psychosocial factors affecting enactment of pro-environmental values by individuals in their organisations. I took a systemic and integrative approach to investigating these factors, through a phenomenological enquiry that situated the research subjects in the dynamics of their work settings and the socio-cultural context in which they and their organisations are embedded. The methodology is transdisciplinary, integrating IPA with frame and metaphor analysis, and draws upon theories of psychological threat and coping, needs, emotion and embodied cognition.

I have provided an example of analytic commentary to demonstrate how the analysis can be written up in a nuanced and in-depth way. The analysis and interpretation includes but also goes beyond a description of how the research subject makes meaning of his experience of emotion. It offers an alternative narrative that explains his emotion regulation in terms of psychological threat and needs satisfaction, enriching our understanding of the experience, conceptualisation and expression of emotions. These insights are important to climate psychology because of the powerful role that emotions play in environmental decision-making and behaviour.

Frame and metaphor analysis uses abductive reasoning, and the analyst has to be extra careful in checking their own biases when drawing inferences from the data. With the integrated methodology presented here, there is perhaps higher potential for idiosyncratic interpretations, making credibility checks a critical feature of the research design. This was the purpose of the final debrief that I held with research participants, where I shared the key findings and sought their feedback. The interpretations offered here are just one possible reading and are certainly not a claim of objective truth. Other theories and approaches would necessarily lead to different findings and interpretations.

There is value in an integrated phenomenological and cognitive linguistics approach to investigate underlying drivers of behaviour, and I hope that other researchers will be inspired to adopt similar approaches. Transdisciplinary research that can engage with the complexity of human–nature relationship in new and creative ways is desperately needed.

References

Andrews, N. (2017a). *Psychosocial factors affecting enactment of pro-environmental values by individuals in their work to influence organizational practices.* Doctoral thesis. Lancaster University, Lancaster, UK.

Andrews, N. (2017b). Psychosocial factors influencing the experience of sustainability professionals. *Sustainability Accounting, Management and Policy Journal, 8*(4), 445–469.

Andrews, N. (2018). How cognitive frames about nature may affect felt sense of nature connectedness. *Ecopsychology Journal, 10*(1), 61–71.

Bion, W. (1962). *Learning from experience.* London: Heinemann.

Clarke, S., & Hoggett, P. (2009). *Researching beneath the surface: Psycho-social research methods in practice*. London: Karnac.

Clayton, S., Swim, J., Steg, J., Devine-Wright, P., Bonnes, M., & Whitmarsh, L. (2016). Expanding the role for psychology in addressing environmental challenges. *American Psychologist, 71*(3), 199–215.

Crompton, T., & Kasser, T. (2009). *Meeting environmental challenges: The role of human identity*. Godalming, UK: WWF-UK.

Deci, E. L., & Ryan, R. M. (2000). The "what" and "why" of goal pursuits: Human needs and the self-determination of behaviour. *Psychological Inquiry, 11*, 227–268.

Deci, E. L., Ryan, R. M., Schultz, P. P., & Niemiec, C. P. (2015). Being aware and functioning fully. In K. W. Brown, J. D. Creswell, & R. M. Ryan (Eds.), *Handbook of mindfulness: Theory, research and practice* (pp. 112–129). New York: The Guilford Press.

Deignan, A. (2005). *Metaphor and corpus linguistics*. Amsterdam: John Benjamins.

Eatough, V., & Smith, J. (2010). Interpretative phenomenological analysis. In C. Willig & W. Stainton-Rogers (Eds.), *The Sage handbook of qualitative research in psychology* (pp. 179–195). London: Sage.

Hasbach, P. H. (2012). Ecotherapy. In P. H. Kahn & P. H. Hasbach (Eds.), *Ecopsychology: Science, totems, and the technological species* (pp. 115–140). Cambridge, MA: MIT Press.

Johnson, M. (1987). *The body in the mind*. Chicago: University of Chicago Press.

Johnson, M., & Lakoff, G. (2002). Why cognitive linguistics requires embodied realism. *Cognitive Linguistics, 13*(3), 245–263.

Kidner, D. W. (2001). *Nature and psyche: Radical environmentalism and the politics of subjectivity*. Albany: State University of New York Press.

Kollmuss, A., & Agyeman, J. (2002). Mind the gap: Why do people act environmentally and what are the barriers to pro-environmental behaviour? *Environmental Education Research, 8*, 239–260.

Lakoff, G. (2010). Why it matters how we frame the environment. *Environmental Communication, 4*, 70–81.

Lakoff, G., & Johnson, M. (1980). *Metaphors we live by*. Chicago: University of Chicago Press.

Lakoff, G., & Johnson M. (1999). *Philosophy in the flesh: The embodied mind and its challenge to Western thought*. New York: Basic Books.

Larkin, M., Watts, S., & Clifton, E. (2006). Giving voice and making sense in interpretative phenomenological analysis. *Qualitative Research in Psychology, 3*, 102–120.

Lertzman, R. (2015). *Environmental melancholia: Psychoanalytic dimensions of engagement*. Hove, East Sussex: Routledge.

Lülfs, R., & Hahn, R. (2014). Sustainable behaviour in the business sphere: A comprehensive overview of the explanatory power of psychological models. *Organization & Environment, 27*, 43–64.

Macy, J. (1993). *World as lover, world as self*. Berkeley: Parallax Press.

Merchant, C. (1983). *The death of nature*. San Francisco: Harper & Row.

Meyerson, D. E., & Scully, M. A. (1995). Tempered radicalism and the politics of ambivalence and change. *Organization Science, 6*, 585–600.

Midgley, M. (2003). *Myths we live by*. London and New York: Routledge.

Plumwood, V. (1993). *Feminism and the mastery of nature*. London: Routledge.

Pryzbylski, A. K., Deci, E. L., Rigby, C. S., & Ryan, R. M. (2014). Competence-impeding electronic games and players' aggressive feelings, thoughts, and behaviors. *Journal of Personality and Social Psychology, 106*, 441–457.

Rogelberg, S. G. (Ed.). (2006). *Encyclopedia of industrial and organizational psychology*. London, UK: Sage.

Rust, M. (2008). Climate on the couch. *Psychotherapy and Politics International, 6*, 157–170.

Ryan, R. M., & Deci, E. L. (2008). From ego depletion to vitality: Theory and findings concerning the facilitation of energy available to the self. *Social and Personality Psychology Compass, 2*, 702–717.

Smith, J., Flowers, P., & Larkin, M. (2009). *Interpretative phenomenological analysis: Theory, method and research*. London: Sage.

Stangor, C. (2010). *Introduction to psychology* (1st Canadian ed.). https://opentextbc.ca/introductiontopsychology/. Accessed 18 July 2018.

Thomas, D. R. (2006). A general inductive approach for analyzing qualitative evaluation data. *American Journal of Evaluation, 27*, 237–246.

Totton, N. (2011). *Wild therapy: Undomesticating inner and outer worlds*. Ross-on-Wye: PCCS Books.

Varela, F., Thompson, E., & Rosch, E. (1993). *The embodied mind*. Cambridge, MA: MIT Press.

Vignoles, V. L. (2011). Identity motives. In S. J. Schwartz, K. Luyckx, & V. L. Vignoles (Eds.), *Handbook of identity theory and research*. New York: Springer.

Weinstein, N., & Ryan, R. M. (2011). A self-determination theory approach to understanding stress incursion and responses. *Stress and Health, 27*(1), 4–17.

Weintrobe, S. (Ed.). (2013). *Engaging with climate change: Psychoanalytic and interdisciplinary perspectives*. Hove, UK: Routledge.

White, L. (1967). The historical roots of our ecological crisis. *Science, 155*(3767), 1203–1207.

Willig, C., & Stainton-Rogers, W. (2010). Introduction. In C. Willig & W. Stainton-Rogers (Eds.), *The Sage handbook of qualitative research in psychology* (pp. 1–13). London: Sage.

Woodward, K. (2015). *Psychosocial studies: An introduction.* Abingdon: Taylor & Francis.

Wright, C., Nyberg, D., & Grant, D. (2012). Hippies on the third floor. *Organization Studies, 33*(11), 1451–1475.

5

Emotional Work as a Necessity: A Psychosocial Analysis of Low-Carbon Energy Collaboration Stories

Rosie Robison

Introduction

In recent decades, the urgent energy and climate change challenges we face have come to be accepted as requiring significant insight and intervention at a human and societal level i.e. the 'technological fix' will not be enough (Sovacool 2014). These so-called 'wicked' problems—which involve many interconnected elements—have prompted greater consideration of how to bring together multi-stakeholder interests and increase collaborative working (Büscher and Sumpf 2017). Different branches of the sustainability-related social sciences are therefore actively exploring questions of collaboration linked to governance (Kemp and Loorbach 2006), city-level partnerships (Bulkeley et al. 2010), and engagement (Pidgeon et al. 2014), amongst other areas. There is thus a clear desire—in both theoretical and practical terms—to understand how to design such collaborations most effectively in order

R. Robison (✉)
Global Sustainability Institute, Anglia Ruskin University, Cambridge, UK
e-mail: rosie.robison@anglia.ac.uk

© The Author(s) 2019
P. Hoggett (ed.), *Climate Psychology*, Studies in the Psychosocial,
https://doi.org/10.1007/978-3-030-11741-2_5

to tackle interlinked sustainability issues in a coherent and long-term manner.

However, the target-driven culture within which such collaborations operate (Robison and Foulds 2018) often pull towards a focus on delivery of outcomes (the 'what') over good collaboration process (the 'how'). Indeed, even when a process is deliberately considered upfront—for example integration of stakeholder engagement and internal communication tools into the delivery of low-carbon projects—the actual 'doing' may still be hard, and at times emotionally intensive, work. By explaining the principles behind and then undertaking a psychosocial analysis of collaboration stories related to meeting low-carbon energy challenges, this chapter explores this emotional work (by which I mean our emotional engagement with others and managing our own emotions). Indeed it is interesting to note that a number of chapters of this book look at different angles of professional work within the sustainability sector and this is a new and exciting extension in the psychosocial field. I further argue this type of psychosocial analysis helps us better recognise the experiences, behaviours and reactions of others as mirroring our own to some degree, disarming unhelpful 'us and them' thinking.

Narrative techniques for data collection and analysis can help us explore this collaborative working for sustainability. They are increasingly being used to explore human and social aspects of sustainability, since they are seen as both offering ways of (i) interpreting *meaning*, and generally going beyond one-way communication of scientific 'facts' (Snowden 2000); as well as (ii) including 'authentic' voices and thereby better understanding different (and potentially marginalised) groups (Ferilli et al. 2016). When undertaken in group settings, narrative methods may also be (iii) a collaborative tool in themselves, bringing together multiple voices (Mourik et al. 2017). Specific to the current chapter, the narrative technique of 'storytelling' has gained in popularity recently, including use with technical (e.g. engineering) communities. Major recent UK and European initiatives include the Energy Biographies project, Task 24 of the International Energy Agency, the Stories of Change project, and a (highly subscribed) special issue on storytelling in the journal *Energy Research & Social Science* in 2016.

However, it is important to remember that consideration of factors *beyond* what people explicitly say or do—such as emotion, affect and intersubjective dynamics, key areas of interest in psychosocial work (Clark and Hoggett 2009)—is not the mainstream in social science sustainability work and is also not inherent in all narrative analyses. Indeed in most cases, storytelling and narrative techniques invoked in energy research do not include any special attention to emotion, with other benefits brought to the fore.[1] Some storytelling projects, however, are finding themes such as emotion, affect and atmosphere to be particularly pertinent (e.g. the Stories of Change project, personal communication with Mel Rohse), and call for these to be more readily acknowledged as a very real part of energy transitions. I discuss these next.

Psychosocial Research, Narratives and Emotional Work

The drastic reductions in carbon footprints being targeted in many countries imply fundamental changes for individuals and societies. However, when challenged to make profound changes to our lives, health and social care researchers and therapists have known for decades that we are very capable of acting in contradictory ways. We may say one thing and do another (ambivalence), hide less desirable aspects of what we think or do from ourself and others (self-deception/defence). Further, when we hold anxieties or fears about change we may mould the information we receive to fit those ideas and justify not changing (denial). We do much of this without being cognitively aware we are doing it, although we can develop ways to recognise these patterns. Far from being surprising 'irrationalities' to criticise others for, these are normal parts of human experience.

---

[1]For example, within the ERSS special issue on storytelling and energy much more prominent frameworks (than emotion) included imaginaries, participatory engagement/vision building, the Multi-Level Perspective, media analysis (including discourse analysis and framing), and social practices.

Narrative Psychosocial Methods and Analysis

How then can we include consideration in our analyses of these elements which are often left unspoken or hidden from even ourselves? This is a central question in psychosocial research.

In the psychosocial tradition, psychosocial narrative enquiry may be used in particular to enable unconscious elements to surface. Methods (and accompanying analysis procedures) which focus explicitly on 'narrative' include the Biographical Narrative Interpretive Method (BNIM) as developed by Wengraf (2001), and the Free-Association Narrative Interviews of Hollway and Jefferson (2000). Their proponents describe these methods variously as 'provocation[s] of storytelling' (Wengraf 2001, p. 111) or aimed at 'eliciting stories' (Hollway and Jefferson 2000, p. 35) and similarly in this chapter I use the terms narrative and story fairly interchangeably. It is worth remembering that not all qualitative research interventions explicitly invite stories. Asking 'how can we engage people with climate change?' for example does not invite a story in the way that 'can you tell me about a time when you discussed climate change with a friend of family member?' does—however participants may and often do respond to all types of questions in a narrative form.

I outline here three core ways in which psychosocial *analyses* of narratives may attempt to go 'beneath the surface', and in this way differ from some other types of qualitative analysis:

1. *Listening to the told story*: Psychosocial analyses of storytelling seek to directly attune to *how* stories are told, and explore what this reveals about emotional position. When we deliberately focus on listening to the 'told story' (i.e. as told by that particular person, at that time), the range of possible responses also expands. Rather than confronting the content of what is said in an interview setting ('why did you do that?'), we could choose to respond more to the underlying tone ('it sounds like you feel...'); when underlying conflicts are present this may be more effective in helping the conversation move forward.

This kind of technique has been developed in both addiction therapies and climate change work, termed 'rolling with the resistance' by Miller and Rollnick (2003). Throughout this process, the listener is recognised as an active part; if a speaker senses they are not being listened to this will directly impact on the narrative (Wengraf 2001) and the researcher's human response to the story also feeds directly into the analysis process. Listening to others' stories, and picking up on emotional cues of the teller either verbally or non-verbally, also involves reflexivity about our own emotions, attitudes and behaviours.

2. *Revealing unconscious defences through stories*: The 'defended' subject (Hollway and Jefferson 2000) and the related theory of the organisational or social defence (see a useful explanation in Mnguni 2010) describe our capacity to psychologically protect ourself (unconsciously) from anxieties through strategies which, when looked at in the abstract, may seem irrational. Such strategies, which a psychosocial analysis will seek to recognise, include: splitting and projection—creating caricatures where certain people/structures/groups/times are seen as wholly good or wholly bad, and can thus be blamed or held up; interpretation of new information in a selective way (including 'forgetting' information) to fit our existing beliefs; holding on to contradictory (i.e. ambivalent) ideas for example related to what we 'deserve' (or what society owes us) whilst overlooking our own impact on others.

3. *The framing of stories in society*: Further aspects of narrative/storytelling analysis concern the framing of narratives and how stories can be used to reinforce or challenge the status quo (Little and Froggett 2009). This present study was not looking at stories created or told in the public domain, however the stories were arguably a means of 'safe challenge' to the status quo in a group-working environment, and indeed this is often advocated as a strength of the method (Mourik et al. 2017). An important observation (from psychosocial research and elsewhere) concerns the dominance of a linear 'problem-action-solution' structure to the stories propagated in very many

social contexts, notably in justifications of political decisions. Little and Froggett (2009) align this with a strong focus on success and failure in many modern cultures. This is a clear example that the way we, as individuals, tell our stories is not simply a product of our 'rationality' (or lack of it) but is also of social context, and this is also something a psychosocial analysis aims to be sensitive to.

Emotional Work

Given how narrative, story-based research may help uncover emotional elements of experience as described above, they seem a fruitful approach to explore emotional work. Emotional work or emotion work (Hochschild 1983) is a fairly broad term which encompasses the management our own and others' emotions. Within this, emotional labour refers more specifically to actively modifying or managing the presentation of one's emotions, for example a nurse taking the time to listen and empathise with a patient when they themselves are dealing with a difficult personal issue. Most notably, emotional work and emotional labour has been explored extensively for health workers and caring professions (e.g. Angus and Greenberg 2011; Ogińska-Bulik 2005) however it has been explored in other contexts including socio-emotional learning aspects of education (Blackmore 1996) and organisational studies (Lawrence et al. 2014).

Whilst there is not an extensive literature looking at emotional and affective aspects of collaborative *sustainability* work, a number of innovative PhDs have begun to demonstrate their relevance through in-depth working with organisations. Thus Mnguni (2008) directly explored the 'psychodynamics of collaborating for sustainability' (p. 39) between four institutes working together in Melbourne to deliver education for sustainability programmes. She writes that she 'seek[s] to alert people working in the domain [of sustainability] to some of the practices that, while appearing rational on the surface, may in fact be counter-productive' (p. 41) looking at defensive aspects. Mnguni asks how we can develop a more 'mature relatedness' and be able to work through tensions, rather than aspiring to entirely smooth collaborations and then have things fall apart when that fantasy is broken. In another doctoral study Reger (2017) undertook interviews with volunteers across 30 community energy schemes in the UK, and highlighted that 'the emotions involved in running CESs should not be omitted from academic discussion' (p. 241). She discussed the fears and anxieties participants held about ensuring they delivered on their parts of projects (particularly which may involve friends and have material implications for others) as well as the sense of enjoyment, pride and building relationships being core parts of the reasons for their involvement.

Methods: Story Collection and Psychosocial Analysis

The stories discussed in the present chapter focussed on collaborative working in low-energy projects. They were created at a workshop run in November 2017 in Cambridge as part of the SHAPE ENERGY (Social Sciences and Humanities for Advancing Policy in European Energy) project—see Robison et al. (2018) for details of the full workshop series across 17 European cities. Each workshop explored local energy challenges, and the Cambridge event aimed to *'bring together invited stakeholders concerned with the future of housing in Cambridge to discuss local targets, and pathways to meet these'* including *'identify[ing] where multi-stakeholder collaborations for local low-energy housing initiatives have been most productive/challenging, and discuss some of the reasons for this'*. There were 28 participants from local policy, local business, community groups and NGOs and universities.

The Stories

The SHAPE ENERGY 'flavour' of storytelling needed to be replicable across multiple contexts and led by novice facilitators in some cases. To this end, a full storytelling facilitation guide (and 2-day training event) had been produced in the previous months (Mourik et al. 2017). In particular, the main storytelling elements were fairly structured.

Thus in the Cambridge workshop, the first storytelling session invited participants to write down a personal experience related to collaborative working (in most cases in low-energy housing projects) which was then shared in small groups. Prepared templates for this suggested a 'beginning/middle/end' structure. The second storytelling session used a 'story spine', and groups worked together to imagine 'visions' which responded to challenges they'd identified from the individual stories earlier. A story spine is a set of beginnings of sentences which can then be completed. In this case, they ran along the following lines: 'In order to respond to *[issue from earlier session]…*', 'between 2018 and 2022, the following projects were run…', 'However, problems/conflicts arose …', 'And particular groups worked to address these by…'. Examples of two of the collaborative stories in full can be found in Robison et al. (2018).

Accompanying Materials

Data gathered on the day included: 27 stories from individuals (one co-facilitator did not write down a story; I wrote one of the 27) plus notes written by participants related to the stories they heard from others; 4 collaborative sets of stories (these had been written up by myself from detailed flipchart notes made on the day, and sent to the group facilitators for their final input); videos of a few attendees telling their 'stories'; participant observation notes of the plenary session taken by myself and another member of the SHAPE ENERGY consortium; reflective notes I wrote in the evening immediately after the workshop; and brief emailed reflections from 6 other co-facilitators on their group work. Also included were two internal project reports (one descriptive—what happened, one reflexive—how it happened) which I wrote in the two weeks after the workshop.

The Analysis Process

For the present chapter I went over the above material several times, mindful of the following analytic tools as identified through the psychosocial review work outlined earlier: (i) the overall emotional picture the stories conveyed (how they were told), (ii) potential defences they revealed, and (iii) how the stories reproduced or challenged the wider status quo of the sustainability professional. I was trying to 'listen' to the participants and often pictured the person concerned (which also involved remembering non-verbal impressions from my extensive interactions with participants before, during, and in some cases after the event), and I deliberately identified the main impression each story made on me. I asked myself questions like: 'how would I feel in that situation?', 'what do they seem to be finding hard?', as well as 'what might I say to this person, to tap into how they feel, and keep the conversation going?' In a few cases I built on this latter question to exchange short follow-ups with participants (primarily via email), which also fed into the analysis. Overall, this process facilitated greater consideration of the *whole* stories, rather than solely focussing on small segments of text (which thematic content analysis often pulls us towards).

Findings and Discussion

In the descriptive internal project workshop report (mentioned in the *Accompanying materials* subsection) I highlighted that the following elements for 'successful collaborative working' were identified repeatedly by participants:

- Multi-stakeholder buy-in early on, and sufficient resources in the set-up phase for detailed data gathering
- Developing a shared vision and managing differences in vision
- Generating agreed metrics and clear mechanisms for sharing data
- Considering the many interconnected elements of sustainability ('joined up thinking')
- Utilising educational methods and/or engagement with other groups.

However, results from this type of 'tell it like it is' thematic analysis (Hollway and Jefferson 2000, p. 56) do not necessarily teach us much about *how* to achieve these outcomes, or what to do if (when) things don't go smoothly. Further, it reflects more what participants profess to believe is important rather than what participants actually prioritise and do when faced with real situations (cf. discussions of 'espoused theory' vs 'theory in use' from Argyris [1976]).

In particular, it is noteworthy that this initial analysis did not raise the question of emotional work *at all*, precisely because it is not often conveyed explicitly in what people say. This is particularly true in a setting where people are presenting their professional selves (ironically thus also undertaking emotional work by presenting or repressing certain parts of the self): yet this is precisely the setting in which research and policy agendas are so often constructed. As we shall see, the following sections uncover important elements of collaborative project work which are thus very often missed out of such conversations.

'Beneath the Surface' Analysis Part 1: Unspoken Emotional Work

In this section, I discuss four emotional work 'tasks' in collaborative projects which were invoked: (1) stepping outside our own defensive boundaries, (2) empathising despite stress, (3) continuing in the face of a lack of acknowledgement, and (4) containing anxieties. These tasks share a commonality in that they are rarely spoken about and may not even be directly recognised by the person doing them.

First then, a repeated sentiment within the stories was the importance of seeing the bigger picture, and some frustration at others' inability to do this:

> No-one achieved what they had wanted to. More discussion at the start would have helped, with a real awareness of different values & needs from the project.

> [it was] sparked by a comment.. '[there is] not the mentality in Cambridge or funding to achieve' quality design of green infrastructure – lack of joined up thinking

However, at the same time as projecting these limitations onto others, I got a strong sense of participants giving signals or hints about the limits of *their own* sphere of influence. It was contained in the way job roles were defined, responsibilities were described, and a sometimes resigned attitude to external factors. In my notes, I indicated how participants' *'professional identity allows discussion of projects in a distanced way'*, so they themselves did not always even appear as part of the story. When I read one story, for example, developments which I knew to be highly distressing for those involved and leading to job losses in some sectors (I used the words 'toxic' in my own notes) were reported in a very measured way such as *'inconsistent policy environment'* or *'original timescales too short'*.

Drawing clear boundaries around what is our responsibility and the options open to us can help 'defend' us, and past work in organisational studies has explored how managers in similar roles and situations may perceive very different sets of choice options (Stewart 1982). But these defences can be problematic when there is a need to 'step up' to deal with new or unexpected tasks, which is what participants were explicitly calling on others to do. Indeed, a key question in psychosocial work related to climate change is arguably how we can take greater 'ownership' of problems, and therefore of solutions, without feeling defeated by them. The collaborative visioning stories, which had fewer constraints (since no one actually had to commit to do them), often aspired to get lots of organisations involved, go the extra mile, and work together for some higher purpose. When we step outside our defensive boundaries in this way we may be 'putting ourselves out there' both emotionally and professionally by saying we are willing to take some extra responsibility, and self-care may be particularly important to avoid issues such as burnout.

Linked to this, it was clear that a time-pressured or stressed environment is perhaps the surest way to ensure people retreat into their own perceived responsibilities, rather than consider stepping outside them:

> They were all quite aware, I felt, of the different values, but they were victims of the process and its timing as much as of anything else.

> Due to project pressure ... I felt project members sometimes took the tenants for granted, or made assumptions.

Here is described an ambivalence: both an awareness at a cognitive level (e.g. of the different values of different groups), but at the same time an inability to incorporate this into future actions. Taking the time to understand or empathise with others despite project pressures involves emotional effort, since we may have to set aside our own anxieties about getting the work completed, having to change our own plans, and so forth. Stress directly impacts on our ability to do this, and organisational norms (where deadlines and targets take precedent) are also designed to discourage this.

A third element was related to a lack of acknowledgement and not feeling valued. In one of the stories told within the group I was facilitating, the teller gave a strong sense of resignation that a lot of the effort put into developing a very low-carbon housing development was not perhaps, in the end, valued by the occupants:

> What 'value' is placed on low carbon homes? – not sure in social housing it is recognised by end user.

It was clear that the response of the users mattered to this person, possibly more than that of colleagues or bosses. This type of lack of acknowledgement can be quite hard to bear, and similarly felt like a situation where a defensive boundary may easily be put up: why— as a highly capable person—would you expose yourself to working in areas where you do not get recognition for your efforts? Interestingly, I observed two related but distinct responses to this type of situation. The first involved exploring what those stakeholders *would* value instead, and moving focus to those areas (thus potentially adjusting the overall 'vision'). The second, in contrast, involved identifying which benefits of the existing vision were priorities for the target group. Thus, in the latter case, one participant had chosen to focus on the goodwill there was towards non-carbon saving related benefits of the project they were leading:

> For some it was more the social angle, fuel poverty, rather than carbon reduction.

This strategy had involved working to contain the anxieties (noted here as worries) of other groups:

> there were some involved who initially could not see the benefits of the project or who were worried that implementing the guidance would create more work … [but] … there was a very clear desire within the organisation to deliver healthy homes that have a positive impact on residents' health and wellbeing.

Another participant saw this type of emotional management of anxieties as critical:

> [Managing people's emotional responses] is a top three priority, **probably the most important priority in many projects** when dealing with a diverse group from different sectors… People who represent an organisation/area/group of stakeholders can feel under immense pressure when comparing themselves to others in a group situation. I have to consider what their primary preference or need is and navigate a solution that includes it or at least a manageable and acceptable way to address it. (bold added)

Further, another participant—when asked—readily listed over 20 specific 'fears' they recognised as held by different stakeholders they worked with, linked to e.g. being seen as fair, professional reputation, future contracts and financial implications, public scrutiny, etc. To cope with others' anxieties about things which had not yet happened, participants had to demonstrate their own commitment in actions as well as words. They had to be a reassuring and constant presence, especially in the face of changing external circumstances (e.g. policy changes) or changes in personnel.

None of this emotional work can be undertaken without support, and emotionally effective teams members recognised this and felt able to both ask for and give such support; this involves being willing to reveal some degree of vulnerability, trusting this will not be exploited. Again I note that this type of skillset was not mentioned as an explicit priority although it was hinted at in terms of good and flexible leadership, which one group particularly focussed on.

'Beneath the Surface' Analysis Part 2: Constructive Challenge

In this section, I talk about emotional work which may be more directly apparent than that discussed in the previous section, and an issue often discussed at the workshop: how to deal with challenge (or conflict) within teams as well as externally. After highlighting the disconnect there can be between the outward impression of a project and actually 'living' it, I discuss a strategy commonly advocated to avoid challenge (i) early vision building, before moving on to one which was observed (ii) external blame, before finally discussing (iii) creating a strong underlying purpose whilst remaining open to changing course.

Recognition of achievements and having some sense of key measures of 'success' helps when one is working to keep a team all pulling in more or less the same direction. Many participants, therefore, had key facts and figures at their fingertips, demonstrating the scale and/or success of their projects:

444 one-hour tours were made by 207 visitors to 13 houses

It is a current project and going well, the time to identify & engage nearly 300 stakeholders

...some 100 schools took part. With tens of thousands of bulbs sold.

Success measures were also evident in more qualitative (rather than quantitative) ways in how people talked about the competency of their teams and the credentials of their contacts; all of this was not done to 'show off' but rather to build legitimacy.

However, this presentation of positive impacts, of course, does not capture everything going on for people actually working within a team. In some stories the contrast between the 'external face' of the project, and the reality of working within it, was brought to the fore:

Everyone over-committed and worked too many hours ... the construction company was difficult to work with, ended up being sued ...

and produced houses that the occupants had lots of complaints with…
Externally facing, the project ironically received multiple awards.

Indeed, in the reverse situation, some implied the disheartening impact
of missing targets even when the team had fought hard to maintain the
overall vision:

> Deciding what is a reasonable reason to miss the target … was challeng-
> ing and somewhat subjective… It is important to have the flexibility to
> allow for failures where intent has been made to hit the target.

Few of the stories presented by participants were 'plain sailing'; indeed
the ones that came closest to this were those where the teller gave the
least away about their own feelings regarding the project. A key point
here then is that there is a tension between presenting a positive face,
and yet acknowledging that there may be times when constructive challenge (management of conflict, changes of plan, etc.) within a team is
needed. How a project deals with disagreement and change is perhaps
as or even more important than its list of 'achievements', in particular when considering the legacy it leaves for future projects including
in terms of capacity building. Indeed the five priorities the participants
themselves explicitly identified (at the start of the *Findings and discussion* section) all essentially call for strategies to manage or engage with
differing views.

This desire is despite (or perhaps because of) the fact that it has
become increasingly hard, in our performance indicator culture, for
organisations to discuss challenge and conflict openly. There is an extensive literature around the damaging effect of policies which discourage institutions from admitting any mistakes or failings at all and how
this may have perverse effects by stifling the very outcomes they seek
to encourage (Hoggett 2010). This can be the case even when framing 'problems' as always having neat 'solutions', which has become the
norm in policymaking and programme implementation, which includes
sustainability initiatives.

How, then, did the storytellers themselves explain that they dealt
with (or aspired to deal with) challenge or disagreement? Here, the

collaborative stories do provide particular food for thought, since we deliberately designed the framework (the story spines) to ask people to imagine problems which might arise and plan for how to deal with these. Indeed reflecting on our own storytelling method design we see that we ourselves replicated the dominant 'problem – action – solution' narrative in our story spines, and although we left space for the ending of the story to be more ambiguous, many did choose to close on a note of resolution.

The most commonly invoked mechanism (highlighted clearly in all four collaborative stories) was stakeholder involvement at an early stage. This was seen as somewhat of a panacea for avoiding all future problems:

> consensus at the beginning is crucial, perhaps through running a 'visioning exercise' at the beginning of projects to make sure everyone is on the same page.

However there was not much discussion of *how* to achieve such consensus, and there seemed to be a great deal of faith that this would set the project on a course to avoid all conflict whereas of course, increased stakeholder engagement does not automatically eliminate objections. In one collaborative story there was more reflective consideration of how this could be problematic or indeed not actually achievable:

> various problems/issues arose, including… how exactly to be inclusive and engage 'everyone' through dialogue.

A second, more defensive strategy, when faced with conflict is displacing blame outside the immediate team (as also highlighted earlier). I have talked elsewhere (Robison and Jansson Boyd 2013) about the 'no blame' approach to sustainability issues. By focussing attention on external issues—the technology, the finance, the timescales, or of course groups outside of those immediately present—we can avoid confrontation and keep good relations with the people who we need to continue working with. This is understandable, however it does not provide a mechanism to hold others to account, in an empathetic way, and thus does not always allow the conversation to move forward. This was clearly recognised by one facilitator when discussing the group work:

> ... we began pointing fingers at particular actors for their responsibility in creating the transition challenge. We first pointed to policy-makers... then tenants and households looking to buy a home, blaming them for not demanding sustainable refurbishments and housing design respectively. Then we began blaming the landlords.... The process continued. Going around in circles in the blame game prevented us from moving forward in imagining concrete projects that could be carried out.

Whilst much has been written about 'in-group and out-group' psychology, blame can also be a channel for our own feelings of shame at not having been able to avoid current challenges. When we think we have failed in some way, we are likely to be particularly exasperated by others' failings (projection) to detract from our own shortcomings.

Aside from perfect agreement from the start of a project, or displacing blame onto other parties, a more constructive strategy involved taking strength from developing an underlying purpose. When I dug deeper into this through a follow-up conversation with one participant, it seemed that it was not so much about a consultation process leading to a compromise which suited all, but rather developing a degree of trust in their own underlying purpose:

> It was just a case of... [pause] reinforcing WHY you're doing it. These are our residents, we are responsible for their health and wellbeing. Getting these arguments clear ahead of time. And if this increases the initial work upfront, to focus on the outcomes... I did get a bit 'short' with some people [who were creating barriers] ... We were lucky because the impetus came from [senior staff who were] able to quieten some of the negative language. By saying, for example 'for goodness sake, stop worrying!' which is something I wouldn't dare to say.

At first reading, this reflection might seem to be a dismissal of emotional issues ('stop worrying!'), however, when related to me the sense this story gave was of that senior staff member being highly supportive of their colleagues, and not being afraid to call out excuses for lack of action. In addition, it is yet another acknowledgement that anxiety can be a major stumbling block in projects.

Constructively challenging decisions and actions is notoriously difficult, and direct challenge—trying to 'reason' people into different courses of action—which can be the first reaction may not be the most effective route if underlying anxieties are present. Within the psychosocial design of the Carbon Conversations group work, Randall (2015, p. 8) advocates that *'empathising with the feeling while at the same time not accepting the excuse often helps people unfreeze their old patterns and experiment with new ones'*. Within a collaborative project—in order to avoid people simply dismissing us as having our own 'agenda'—we also need to be open to allowing challenge *to ourselves*. Self-esteem (in part generated through emotional support from others) allows us to contain defensive reactions when challenged. I would also argue that the very type of analysis conducted here helps us avoid 'us and them' defensive thinking through highlighting the commonalities of our emotional experience.

The accounts also show how our approaches to conflict are hugely bound up in our professional practices and structures, for example, the seniority of staff making a tangible difference to what they were allowed to say. Although several of our participants were operating in a relatively independent or consultant capacity, all had to fit within the larger system, and thus negotiate between acting as they are expected to within that system, and asserting their personal feelings of what is desirable. We may be aware of examples when professional expectations (e.g. flying to project meetings) conflict with the messages we are advocating, yet feel unavoidable. Again, the balance between maintaining underlying purpose and openness to other viewpoints is a critical one.

Conclusions: Emotional Work as a Necessity

Much of what I have discussed could relate to collaborations in many contexts, however, these are important issues to highlight in the low-carbon or sustainable energy context in part since energy (and climate change) have long been seen as primarily technical domains. This means emotion is often overlooked, whereas it is more taken for granted in, for example, health work. When using a psychosocial analytic approach with these sustainability-related collaboration stories, skilled emotional work is seen as a necessary part, which often involves navigating fine balances between our own and others' emotions. Listening to the 'told' stories and recognising defences offered significantly more detail than a content analysis in terms of the everyday actions needed in order to actually manage, for example, 'differences in vision'. These included recognising others' fears, acknowledging the efforts of others, and noting how stress can limit empathetic response. Considering the stories' framings we also directly saw the tensions created through the need to adopt a somewhat falsely positive external face to projects.

A critical point, of relevance for future studies, is that although emotional work was not identified upfront by participants (and thus would not be readily developed from a thematic content analysis), when reflected back to participants (through sending follow-up questions) its relevance *was* directly recognised. This reemphasises both how methodologies themselves determine what types of results can be found, and how we are still in the early days of awareness over the importance of emotional management. How do these results practically help us? Not all sustainability professions will want to (or should) suddenly pursue a professional accreditation in emotional work, but it arguably should be given greater prominence in designing projects. Expertise in this domain is not always included in job descriptions or recognised professionally. It is not an aim of low-carbon energy policies to help develop emotional skills, and some may even experience it being actively defended against in some settings.

One criticism of these in-depth qualitative techniques is their lack of scalability, given the enormity (and urgency) of the challenges faced. Psychosocial researchers may for example conduct multiple interviews with the same person over several months. This chapter therefore also offers a contribution in that it reports on the use of a method which has been 'scaled' to a certain degree. The stories analysed were prompted using a relatively structured approach, and the amount of time given to their telling was not extensive. Did this attempt at scaling work? Gathering data in a narrative format has certainly allowed a far deeper post-event analysis related to the feel and emotion of the responses, but also facilitated more nuanced and reflective discussion between participants at the workshop itself, particularly for the collaborative visioning stories. The psychosocial analysis undertaken for this chapter was however only made possible through my personal engagement and dialogue with the participants, something that more conventional large-scale methodologies prohibits.

The results suggest that—rather than placing faith in building perfect consensus at the start of a project—it may be a combination of hundreds of small actions and decisions (to acknowledge, to empathise, to recognise both others' and our own excuses, to seek support, to press on) which add up to a collaboration which more successfully deals with this emotional work. These small actions are rarely captured in an interview or one-off setting and diary work (coupled with some form of personal interaction) performed by those undertaking active projects could be a fruitful next step. A final challenge is that in all of this, to be successful, emotional support must feel genuine rather than something undertaken to tick the right box. Whilst creating compulsory new 'targets' ('you *will* be emotionally supportive to colleagues') may, therefore, be counterproductive, it *is* possible for institutions to foster cultures where it is ok to show this vulnerability and accept support, and this must be a greater part of the conversation around low-carbon energy transitions.

Acknowledgements The research which fed into this chapter was funded through the SHAPE ENERGY project, as part of the European Union's Horizon 2020 research and innovation programme under grant agreement No. 731264. My sincere thanks to all participants of the workshop for their time and insights; if you read this chapter I hope you find it of interest. Thanks also to Jo Hamilton, Paul Hoggett, Rosemary Randall and Mel Rohse for comments on earlier drafts of this chapter.

References

Angus, L. E., & Greenberg, L. S. (2011). *Working with narrative in emotion-focused therapy: Changing stories, healing lives*. Washington, DC: American Psychological Association.

Argyris, C. (1976). Single-loop and double-loop models in research on decision making. *Administrative Science Quarterly, 21*(3), 363–375.

Blackmore, J. (1996). Doing 'emotional labour' in the education market place: Stories from the field of women in management. *Discourse: Studies in the Cultural Politics of Education, 17*(3), 337–349.

Bulkeley, H., Broto, V. C., Hodson, M., & Marvin, S. (Eds.). (2010). *Cities and low carbon transitions.* London: Routledge.

Büscher, C., & Sumpf, P. (2017). *Energy & multi-stakeholder interests—A social sciences and humanities cross-cutting theme report.* Cambridge: SHAPE ENERGY.

Clarke, S., & Hoggett, P. (2009). *Researching beneath the surface: Psycho-social research methods in practice.* London: Karnac Books.

Ferilli, G., Sacco, P. L., & Blessi, G. T. (2016). Beyond the rhetoric of participation: New challenges and prospects for inclusive urban regeneration. *City, Culture and Society, 7*(2), 95–100.

Hochschild, A. R. (1983). *The managed heart: Commercialization of human feeling.* Berkeley: University of California Press.

Hoggett, P. (2010). Government and the perverse social defence. *British Journal of Psychotherapy, 26*(2), 202–212.

Hollway, W., & Jefferson, T. (2000). *Doing qualitative research differently: Free association, narrative and the interview method.* London: Sage.

Kemp, R., & Loorbach, D. (2006). Transition management: A reflexive governance approach. In J. Voß, D. Bauknecht, & R. Kemp (Eds.), *Reflexive governance for sustainable development* (pp. 103–130). Cheltenham, UK and Northampton, MA, USA: Edward Elgar.

Lawrence, E., Ruppel, C. P., & Tworoger, L. C. (2014). The emotions and cognitions during organizational change: The importance of the emotional work for leaders. *Journal of Organizational Culture, Communications and Conflict, 18*(1), 257–274.

Little, R. M., & Froggett, L. (2009). Making meaning in muddy waters: Representing complexity through community based storytelling. *Community Development Journal, 45*(4), 458–473.

Miller, W., & Rollnick, S. (2003). Motivational interviewing: Preparing people for change. *Journal for Healthcare Quality, 25*(3), 46.

Mnguni, P. P. (2008). *Mutuality, reciprocity and mature relatedness: A psychodynamic perspective on sustainability.* Unpublished doctoral dissertation, Australian Graduate School of Entrepreneurship, Swinburne University of Technology.

Mnguni, P. P. (2010). Anxiety and defense in sustainability. *Psychoanalysis, Culture & Society, 15*(2), 117–135.

Mourik, R., Robison, R., & Breukers, S. (2017). *Storytelling—SHAPE ENERGY facilitation guidelines for interdisciplinary and multi-stakeholder processes.* Cambridge: SHAPE ENERGY.

Ogińska-Bulik, N. (2005). Emotional intelligence in the workplace: Exploring Its effects on occupational stress and health outcomes in human service workers. *International Journal of Occupational Medicine and Environmental Health, 18*(2), 167–175.

Pidgeon, N., Demski, C., Butler, C., Parkhill, K., & Spence, A. (2014). Creating a national citizen engagement process for energy policy. *Proceedings of the National Academy of Sciences, 111*(Suppl. 4), 13606–13613.

Randall, R. (2015). *The carbon conversations facilitator's guide.* Stirling: The Surefoot Effect.

Reger, C. E. S. (2017). *Understanding the post-start-up experiences of community-based energy schemes in the UK.* Unpublished doctoral dissertation, University of Sheffield.

Robison, R., Dupas, S., Mourik, R., Torres, M., & Milroy, E. (2018). *Europe's local energy challenges: Stories and research priorities from 17 multi-stakeholder city workshops.* Cambridge: SHAPE ENERGY.

Robison, R. A. V., & Foulds, C. (2018). Constructing policy success for UK energy feedback. *Building Research & Information, 46*(3), 316–331.

Robison, R. A. V., & Jansson-Boyd, C. V. (2013). Perspectives on sustainability: Exploring the views of tenants in supported social housing. *Sustainability, 5*(12), 5249–5271.

Snowden, D. (2000). Storytelling and other organic tools for chief knowledge officers and chief learning officers. In D. Bonner (Ed.), *Leading knowledge management and learning* (pp. 237–252). Alexandria, VA: American Society for Training and Development.

Sovacool, B. K. (2014). Energy studies need social science. *Nature, 511*(7511), 529.

Stewart, R. (1982). A model for understanding managerial jobs and behavior. *Academy of Management Review, 7*(1), 7–13.

Wengraf, T. (2001). *Qualitative research interviewing: Biographic narrative and semi-structured methods.* London: Sage.

6

Researching Climate Engagement: Collaborative Conversations and Consciousness Change

Sally Gillespie

Introduction

Depth psychological research proceeds from the hypothesis that climate disruption, and reports about it, arouses intense and often unbearable feelings, including guilt, fear grief, and despair, which challenge people's ability to stay engaged with the issue. This makes questions about the psychological life and processes of people who do maintain an ongoing engagement with climate issues of great interest. Such people are the outliers, the canaries in the mine, who can report back on their experiences of living with climate consciousness. They are also well-positioned to reflect on unconscious responses to climate disruption, when supported to do so, due to their experience of consciously negotiating terrain that remains largely unconscious for most people.

Identification of unconscious dynamics is crucial for increased understanding of how climate change affects us psychologically. Depth psychology is particularly well-equipped to analyse and work with these

S. Gillespie (✉)
Sydney, NSW, Australia

© The Author(s) 2019
P. Hoggett (ed.), *Climate Psychology*, Studies in the Psychosocial,
https://doi.org/10.1007/978-3-030-11741-2_6

responses because its primary focus is on the interplay between conscious and unconscious dynamics within individuals (intrapsychic) and between people (interpsychic). Furthermore, depth psychological practices support the creative and healing possibilities of making what is unconscious, conscious. These practices work through eliciting reflections, feelings, body sensations, images and insights within a framework of attentive listening. This approach fits well with qualitative research and a phenomenological approach which both focus on complex and rich descriptions of lived experience.

In order to elucidate the many layers of consciousness and unconsciousness at play in ourselves and the world, depth psychological research employs a practice of "careful attention that is sustained, patient, subtly attuned to images and metaphors, tracking both hidden meanings and surface presentations" (Watkins 2008, p. 419). This attention supports analyses grounded in close observation of reports of subjective knowledge and experience which focus on "being with" through reflection and dialogue, while "seeing through" the apparent to the less apparent (Hillman 1992, pp. 140–145). Nevertheless identifying what is unconscious is a notoriously tricky and complex business, especially when the researcher's own unconscious dynamics are taken into account, as they must be within a depth psychological framework. Researchers need to be trained to listen deeply to others within a form of open enquiry, which pays attention to what is not said, as much as to what is said.

Symbolic language, memories, fantasies, feeling states, dreams and somatic responses are all data in depth psychological research. Dreamwork, in particular, can explore the creative potentials of what is habitually unconscious, opening up possibilities for new perspectives and directions. Carl Jung (1974) believed that "dreams are the most common and most normal expression of the unconscious psyche, [providing] the bulk of the material for its investigation" (p. 73). Dreams frequently confront us with what we find too discomforting, frightening, or even awe-inducing to fully face in waking life. The inclusion of dreams and dreamwork in climate psychology research helps identify, articulate and validate unconscious responses to climate disruptions and/or the research process that would otherwise be neglected, avoided or repressed.

Depth psychological research into climate engagement investigates possible outcomes when people are supported in exploring their awareness about climate disruption and their involvement with it (Randall 2012). The research discussed in this chapter explored the moral, existential and psychological dilemmas provoked by awareness of climate change through ongoing discussions in a group of participants engaged with climate issues. Its phenomenological orientation encouraged an experiential focus which helped participants to cultivate awareness of their ways of being in the world, helping to make "it a kind of therapy for healing the splits of modern thought" (Fisher 2013, p. 11). This therapeutic benefit was further strengthened by paying attention to participants' dreams. While the research process necessitated the acknowledgment of many disturbing feelings and experiences, participants reported increased feelings of resilience, maturity, compassion, motivation and commitment to climate action at the end of the year-long process.

Collaborative Research

Climate disruption is a collective phenomenon requiring systemic responses. These include psychological research projects which draw upon shared frameworks of reflection. Collective research offers possibilities for collaborative knowledge making, while facilitating dialogues that support psycho-social developmental processes and resilient engagement (Gillespie 2013). For these reasons, I conducted my research through a participatory action research group which met twelve times over the course of a year.

Critical participatory action research is carried out by collaborative groups who generate their own questions, critical understandings and informed actions by paying attention to their psychological experiences, alongside the social and historical contexts of their situation. Depth psychologists Mary Watkins and Helene Shulman (2008) advocate the use of critical participatory action research to address collective traumas inflicted by colonialism and globalisation. This category of traumatisation could well include climate disruption (Klein 2014). Participatory research encourages participants to articulate multiple perspectives, while supporting their efforts to "break out of closed systems of symptom and isolation" (Watkins and Shulman, p. 277). It challenges individualist paradigms of research, just as climate disruption itself challenges individualist conceptions of identity and action. Esther Madriz (2003) observed that "In a culture that highlights individualism and separation, shifting the research agenda in the direction of commonality and togetherness is, in itself, subversive" (p. 373). Within this climate psychology research, a subversion of beliefs in individualism and separateness contributed to the development of collaborative views and practices. These practices, in turn, supported the development of ecological consciousness, with its validation of a multiplicity of interacting viewpoints and experiences.

Participatory research aims to build knowledge through mutual revelations, observations, questions and interactions that can extend the boundaries of personal consciousness. This doctoral research, based in Sydney, Australia, stemmed from a desire to understand the psychological effects of my own and others' immersion in climate change issues. It followed several years of my own reading, dream journaling and writing about this issue, based on my interests as a psychotherapist and fears as a human being. I found research participants by asking friends and colleagues to send out my invitation through any general mailing list they were on. Seven people (all but one unknown to me) volunteered to join me for a series of twelve two-hour group discussions to research their subjective experiences and understandings of climate engagement. All lasted the distance (although some missed an occasional meeting). I also interviewed each participant twice, at the beginning and end of the research cycle.

Participants came from a variety of professional backgrounds, including community development, art, energy policy, law and academia, contributing a range of skills, resources and perspectives to the research. All were engaged with climate concerns in some way, ranging from policy making and communications through to activism. The group bonded over shared climate concerns, as well as feelings of social alienation arising out of their immersion in this generally marginalised, if not denied, issue. While many action research projects work with the pressure to come up with new or changed policies or practices at the end of the research cycle, this research group had the spaciousness to focus on the complex and confronting psychological dynamics, that preceded, accompanied or followed climate action. Participants reported feeling both relief and stimulation in their experience of open dialogues with people who worked in a variety of fields, without an expectation of ongoing collaboration, or a fixed target.

Drawing upon co-operative inquiry and participatory research models (Heron 1996; Reason and Bradbury 2001), I viewed myself as both initiating researcher and co-participant, and my co-participants as collaborators who sought data through questions, and provided and analysed data through discussions and reflections. Nevertheless, recognition of collaborative reciprocity did not confer equality of power, responsibility, acknowledgement, comprehension or contribution. It was clear to all that I was the research leader responsible for the conduct of the overall project. My ethical commitment and proficiency as a researcher with psychotherapeutic expertise was viewed as particularly important by my co-participants. They repeatedly acknowledged my work in creating and maintaining a safe space as a crucial factor for their ongoing involvement.

Overall the collaborative nature of the research evolved most successfully through open-ended discussions, which provided data in response to research questions proposed and developed by all the participants. Some initial data analysis was also performed collaboratively in these discussions through questioning one and other's contributions and responses. My invitation to contribute through further data analysis and/or writing after the research cycle ended, was largely not taken up. Participants said they were too busy and/or not interested in performing work which they assumed (correctly) was mine to perform as a doctoral candidate.

The two-hour meetings were held at a community centre. The guidelines I gave were for respectful speech and listening without interruption. Everyone agreed to the discussions being taped and transcribed by me, with transcripts being sent out to all for checking. Both depth psychological research and co-operative inquiry draw upon phenomenology, in their valuing of experiential knowing, and use of open, unstructured discussions as a form of inquiry (Coppin and Nelson 2005; Heron and Reason 2008). While I proposed some research questions in my initial advertisement for co-participants, I stated that they were only a beginning indication of the areas that interested me. I explicitly invited other questions and foci to emerge from the group's discussions, which is what happened. Aware that my preferred orientation, depth psychology, was possibly not shared by other participants, I arrived at our first meeting with suggestions for a more structured approach, using stimulus materials or topics. The group vetoed this approach, with a number of participants expressing the view that they already felt immersed in climate change information. Instead, they welcomed the opportunity for exploratory discussions around their own thoughts and feelings, stories and dreams. One participant, Linda, declared that "the most exciting aspect of what is happening here, is … not what we know already… or opinions we already have. It's about what we don't know, and that we might find another way". After that, my only contributions towards the structuring of the discussions were the time keeping ones of beginning meetings with a simple welcome and winding up at the end. People arrived eager to talk and prompts of any other kind were not needed after the initial two sessions.

The focus on exploring what *"what we don't know"* in order to *"find another way"* became a beacon for this research. Drawing upon my depth psychology experience, I trusted the value of following the group's free-flowing conversations as long as, in the group's consensus, we kept to the agreed overall agenda. The group embraced the opportunity to interrogate and test their beliefs and experiences about advocating for collaborative action to address climate disruption. The need to listen to others, value diversity, tolerate conflicting views, consider overall objectives and develop research protocols and strategies were principles common both to collective action and the action research group's emerging process. The commitment of the group to the focus and form of this inquiry created the safety to question identity, assumptions, values, ways of living and forms of engagement. Being able to explore and negotiate contested knowledge and meanings through discussions produced insight, energy and resilience for climate change communication and action. Another significant effect of the participatory nature of the research was to affirm the value of participatory processes generally, and the presence of collaborative processes of knowledge and change, within human and other than human ecosystems. The group's acknowledgement and exploration of diverse and systemic processes helped to shift worldviews away from the pursuit of single-minded perspectives or hierarchical strategies. As a result, discussions became more robust in their negotiations of the complexity and messiness of ourselves and the world, as participants found that what they and others had habitually censored, marginalised or overlooked could now be explored.

Climate Conversations

Paying attention to those who are engaged with climate issues provides insights into the psychological journey from old to new worldview, as personal and social identities and meanings are reworked. Writing about the transitional psychic space between old myth and new myth, Ginette Paris (2007) observes that it "often feels like a deadly zone" which depth psychology can helpfully illuminate (p. 83). Those who are engaged with climate disruption are at the forefront of a shift from old myths (or worldviews) to new ones. Within this process they commonly encounter primal emotions and concerns about existence, security, identity and place. Deathly imaginings, fears and griefs sit close to the surface, although often suppressed. However, this is only part of the psychological story of climate engagement, as along with survival anxiety, comes potentially transformative opportunities for finding and telling new stories about self, society and world (Connor 2011; Hoggett and Randall 2016).

There is both need and opportunity for those who are engaged with climate disruption to meet with others to reflect in depth on the nature, process and effects of their engagement. In this research, participants were quick to acknowledge and explore a range of horrors and terrors within a confronting worldscape, rife with losses and disruptions. Apocalyptic imaginings and nightmares featured especially strongly in the first three meetings' discussions. Through taking time to explore the form, content, meanings and emotional responses to this material, participants found an opportunity to collectively acknowledge and digest what many people experience as a "no go" zone. For Veronica, this sustained focus spoke "to the unmet need that we have to be with this". She declared that "the greatest gift of this group for me has been a place to show... depression is not the right word, despair is not the right word, but the darkness of the knowledge I hold". Despair and terror are highly challenging feelings to continually live with, physically, mentally and emotionally. Psychologically it is healthy to seek ways to minimise or avoid the traumatising nature of such feelings. In the group's dialogues, expressions of despair were interspersed with ones of hope. This interweaving of pessimistic and optimistic views, both between participants and within participants, helped to steady discussions. It seemed that acknowledging imaginings of future creative or healing developments did not disallow the dreads and fears that all participants entertained at different points of the discussions, but actually helped to explore them. And vice versa. Through repetition and analysis, the group came to understand that each future vision was just one of many possible imaginings that may or may not play out.

Being able to be honest about this emotional seesawing stimulated ongoing discussions about the causes and effects. Over time, these reflections facilitated an informal and spontaneous practice of mindfulness about hopes and fears in relation to climate disruption specifically, and imagined global futures generally. Sara talked about the inevitability of pessimistic times, validating the deeper awareness that this brings:

> It is important to be in... spaces along the continuum... it gives you that appreciation of where people can go, and while I find it very uncomfortable to be in a dark moment... as I've gotten older, I find it easier to be there, and I do remind myself that I'm not going to be there forever... I suppose that's part of living with the complexity and the contradictions, because I feel I have a whole lot of contradictions within me.

Learning to tolerate complexity and uncertainty develops psychological resilience. Psychoanalysts elaborate the necessity of consciously enduring ambivalent feelings in order to psychologically mature. Carl Jung (1969) describes how holding the tension of seemingly opposing viewpoints, dismissing neither, and thus avoiding the destructive and distorting effects of one-sided judgements and positions, can result in a liberating transformation of consciousness. In their research, Paul Hoggett and Rosemary Randall (2016) observed that when activists can hold the tension between optimism and pessimism they have "less need to resort to binary thinking as a way of engaging with reality". In our group's discussions, holding the tension of opposites between hope and despair, facilitated a dynamic process that was unsettling and freeing as it supported a greater consciousness of the world's complexity and human life within it.

One of the most immediate benefits of the group's dialogues was to alleviate the mental and emotional isolation that many people who are engaged with climate issues experience. Through staying present to both personal and public dimensions of their engagement, participants in this research worked to dismantle artificial constructions which distort and diminish personal identity and agency. This not only brought about immediate relief and greater emotional wellbeing, but also the possibility of thinking beyond individual and social constraints towards the larger collective dimensions of life. Developing a systemic consciousness reduced feelings of isolation while supporting people to fundamentally re-vision themselves and their values.

The process of sitting in conversation with the range of dilemmas posed by climate disruption, accepting discomforts and the consciousness of repression, proved to be sufficiently dynamic and broadening to provoke changes in values and behaviour for many in the group. For Sam, one consequence of being a part of the research was to volunteer with organisations engaged with climate issues. He told the group that he hoped that:

> The volunteering I am doing will help to maintain that broadened approach… there is the tendency in myself to go to the more hierarchical individualistic side, and I'm trying to catch myself before I go that way.

The conscious holding of changes such as Sam's through the group's dialogues provided foundations for different meanings and directions to form; ones which contributed to both psychological development and socio-political change, as I explore later in this chapter.

What Lies Beneath

While this research worked through the medium of conversations, its data at times moved from the realm of the verbal into the imaginal and the symbolic. In order to capture something of the depth and breadth of the research group's conversations, I recorded data on a map of an imaginary world.

Continents represented major topics of conversation (Communication, Engagement, Change, Death, Survival, Material Life and Relating to Earth), while geographical features on each continent were labelled using direct phrases from the discussions (e.g. *Switching Off Point, Bamboozled Bay, Lake Inexpressible, Smart Monkeys Rock, Mt Hairshirt, Death Denial Canyon, Cut off Canal, Visceral Beach*). This mapping gave physical reality to the group's discussions. It also worked symbolically to crystallise imaginings of a transforming world, which both generates disrupted understandings and feelings while inviting exploration.

When presenting this research at talks and conferences, I found that this storied "world map" resonated with many. It seemed that the map's form and phrases contained, and to some degree transformed, a *massa confusa* of emotions into an embodied image of newly emerging views and experiences of the world, offering a foothold for some sense of agency. Sara, a research participant, commented that for her the effect of this map was:

huge, because... it summarised what we were saying, or made it concrete, and somehow for me it made it possible. How magical is that?... For me to learn I need to go and stand somewhere where I have never been before, and have that experiential learning through other people.

Symbols evoke a feeling of the "magical" through their numinous and transformative effects. Their presence opens up what cannot consciously be fully perceived or articulated, while energising the desire and ability to undergo deep change. Sara's comment identified the symbolic nature of the map, emphasising its transporting qualities which took her to a place where it was possible to be present to climate disruption through a collective learning experience. For others, this map's symbolism works to contain, inspire and motivate people through its acknowledgement and representation of the magnitude of climate issues within a context of discovery and adventure.

Depth psychological research into climate disruption responses sails deep waters with many hidden currents. Sharing dreams offered one possibility to identify and articulate these unconscious undertows. Nevertheless, my original suggestion to share dreams within the group's discussions was tentative as I did not know if this practice would engage participants. As it turned out, dream sharing became a significant, although not a major, part of research discussions. By the end of the meetings everyone had shared at least one dream, and dreams were discussed at every meeting. At times these discussions were brief, at other times extended. As meetings progressed and participants became more at ease with dreams, their material became woven into discussions, rather than treated as a separate topic. When a dream was shared, the subject and tone of discussions invariably changed in some way. For some, the dream sharing was a major part of the research experience, facilitating unexpectedly powerful shifts. For Linda, it was "a way of fast tracking, getting to a place in a group... in a supportive environment that you wouldn't otherwise do... it made us go to a place that was very primal".

The dreams' symbolic images and narratives became another voice within the research; often unsettling and provocative and frequently moving. In his work with groups of Vietnam veterans, Robert J. Lifton (1996) described how "Dreams seemed to possess something in the order of mythic power, of special illumination, that in turn enabled the veterans – as individuals and as a group – to make their special psychic leaps into the most painful terrain" (p. 130). Similarly the dreams shared in our discussions facilitated expression of vulnerabilities, fears, despairs and griefs that lay outside the boundaries of our customary thoughts and discussions. Listening and responding to these dreams, especially the nightmares, developed group and individual resilience to consciously face and explore worst fears and horrors, and then move on.

Feelings of hopelessness, shame, urgency and terror all became major topics of discussions as a result of dream sharing. One dream shared by Veronica in the fifth meeting opened up the area of environmental activism and trauma. In her dream, two old musician friends who she described as "spiritual journey mates" were discriminated against by the general public. The dream then turned into a nightmare when an orphaned teenage boy who befriended these musician friends had his head sliced in two by a masked man. Veronica told us that she witnessed this in her dream like she was "an angel in heaven...I watch his head fall into two pieces, and as I'm watching I'm just devastated. I... woke up devastated with that horrible, horrible feeling of violence".

The experience of the dreamer as a disembodied witness to horrific acts is a common feature of trauma dreams, demonstrating how the ego disassociates, or isolates itself, from what is experienced as unbearable (Bosnak 2007, p. 65). Veronica had this dream after she attended a particularly disturbing presentation on the dangers of radioactive poisoning in the wake of the Fukushima meltdown. Recounting this to the group, she described horrific possible mutations, barely drawing breath. Her terror and overwhelm were palpable. Hearing her dream and feeling Veronica's continuing distress strongly affected the group. The intersection of individual and collective traumatisation requires acknowledgment and careful handling. When this happens within the context of research, Watkins and Shulman (2008) advise that:

> Careful listening to narratives is particularly important ...It is necessary to listen to what is on the edge of coming into words, for what yet cannot be clearly stated, for the ellipses between assumed certainties, and for the iconic meanings that stand in for experiences that were overwhelming. (p. 287)

In this instance, participants responded to Veronica's dream with care and interest, intuitively sensing that there were meanings within it that could not be immediately articulated. By staying close to the dream experience, Veronica identified her overwhelm in response to hearing about ecological harms in the world. She then could relate to the discrimination against her friends in the dream, as she recognised the marginalisation of spiritual and creative values in both herself and in contemporary culture.

A further discussion evolved when Veronica described the talk she had attended, commenting that there had been a

> very bizarre energy between [D - the presenter] and the college students in the room... Their affect was interpreted by D as not caring and not understanding, and there was a bit of hostility during the talk. I think D is also perhaps at risk of burnout... I asked her, after her talk, 'I can't even quite imagine how to open up my brain to let in this information. I'm afraid it could make me unable to function to have this information... How do you function?' And she... said back to me "what I figured out, if I don't get up every day and do this, I have to go on Zoloft".

Hearing this, several participants made the link between Veronica's question about opening up her brain and the slicing open of the boy's head in her nightmare; a connection which resonated with Veronica. In the symbolic language of the dream, there had been a traumatic onslaught on the mind, which Veronica was now able to consciously recognise and articulate. As a result, the group could consciously grapple with the ways that violence, damage and disconnection can play out within ourselves and the world in relation to environmental activism. This discussion also facilitated an experiential learning process about mindfully monitoring the effects of what we read and hear, as well as bringing consciousness to language and delivery when communicating about climate disruption. Shocking facts delivered without attention to the emotional responses of the audience can be traumatising, defensively increasing responses of denial. Supportive contexts with time to digest bad news and reformulate different understandings about ourselves and the world are crucial to avoid traumatisation, and the risk in turn of unconsciously traumatising others. Equally important for effective and resilient activism is the need for creative and spiritual nourishment and renewal (Hoggett and Randall 2016).

Sharing dreams connected research participants to a resource that was nourishing, stimulating and critically informing at personal and collective levels. While a common view of dreams often rests upon a perceived duality between inner and outer life, dreams themselves support symbolic and metaphoric understandings that connect, rather than divide, inner from outer or personal from collective (Bulkeley 2008; see also the Manley and Hollway contribution to this volume). Many dreams shared in this research facilitated processes of making conscious connections to feelings and attitudes of disconnection that were largely unconscious, thus inviting what had been psychologically exiled into both individual and group consciousness. Significantly two of the most inspiring dreams shared in the meetings both contained powerful images of connection, one of which is described below.

Developmental Possibilities

Groups, like dreams, can cut through habitual perspectives. Beneath the surface presentations of seesawing emotions, the group perceived larger developmental processes at work which transformed worldviews and perceptions of self. Veronica observed that "there's been a maturation process for me going on in this group. And part of that is to go from self to immediate community then to planetary or interplanetary perspective from the short run to the long run". She also identified an increased ability to hold "apparently conflicting information and knowledge" without feeling compelled into some kind of reactive response, and to be with the uncertainties of "the unknown rather than the hyperbole of immaturity which is 'Oh my god, the sky is falling!'".

The group's discussions increased capacity for tolerating complexity and uncertainty as well as for acknowledging, holding and sharing contradictory or uncomfortable feelings. The frequency with which participants observed polarisation occurring within climate debates, stimulated many discussions in the group about the importance of holding and exploring the tension between opposites, thus avoiding the destructive and distorting effects of one-sided judgements and positions. In the third meeting, Lisa shared a dream which vividly portrayed a holding of opposites:

> I dreamt I bore twin babies. Baby boys – one was black, and one was white... Here were these two babies and I was holding them and feeding them... I woke up from this dream... feeling this was a resolution between opposites. And I wasn't even specific in my mind what these extreme opposites were. It was just this wonderful feeling of differences meeting. It was quite profound.

Lisa described her dream as a primal experience in which she could feel, and even smell, the newborn babies in her arms, arousing feelings of both nurturing and being nurtured. While Lisa was aware that there were many things she could read into her dream, what spurred her to share it with the group was her awareness that the group was dealing with extreme views and feelings in its conversations. I suggested that we could sit with this dream as a group, allowing its sensual and powerful imagery, to hold us as much as we might hold it. This approach sat well with the phenomenological orientation of the research as well as the group's preference for responding experientially rather than analytically to dreams that were shared.

Lisa's dream offered the group an embodied and integrated image of holding seeming opposites with an ease of acceptance that transcends discriminatory or divisive judgements. The dream's imagery and feeling tone illuminated and furthered the evolving conscious directions of the research which was to challenge single-minded and potentially conflictual positions. Much of the ongoing work of the group pivoted upon identifying and containing perceived dualities such as optimism and pessimism, life and death, empowerment and disempowerment, individual and collective, present and future. This work forged a creative and dynamic process that embraced paradoxes, contradictions and uncertainties, opening minds and maturing psyches while building a greater consciousness of the world and our place within it. The group's developing ability to recognise the articulation of polarities as a beginning point for integrative understandings, rather than as a fixed point of conflict, provided a strong basis for psychological development and transformative learning processes (Boyd 1991, p. 10).

Research discussions repeatedly circled around the necessity, challenges and opportunities of dialoguing with difference, in order to move beyond the distortions and limitations of singular understandings and views. The group process of opening to a multiplicity of understandings beyond habitual ones was a highly valued experience, frequently mentioned within the participants' final interviews. The group's rewarding experiences of working with a diversity of outlooks and multiple forms of knowing, acknowledging biases while extending understandings, encouraged pursuing such approaches beyond the confines of the research.

An associated psychological developmental task carried out within the group involved withdrawing projections on to others and the world. Projections occur when, as a defence against anxiety, individuals or groups unconsciously transfer challenging emotions, disturbing thoughts or undesired aspects of themselves onto others (Samuels et al. 1986, p. 113). Learning to identify and withdraw projections is vital to maturity. In this research for example, when discussions started to focus on blaming and shaming people or groups out there in the world (typically politicians or corporations), someone in the group would often turn the spotlight on to our own thinking or behaviours, interrogating our unconscious motivations and analysing our responses.

Personal and political consciousness deepens when we can move beyond projections, which work to deflect fault and blame onto another person or group, while excusing ourselves of any shortcomings or responsibilities. In his research into liberation movements, Lawrence Alshuler (2006) analysed points of conjunction between a model of developing psychological consciousness theorised by Carl Jung, known as individuation, and a model of developing political consciousness theorised by Paulo Freire, known as conscientisation. The goal of both the individuation and conscientisation developmental models is to shift limited personal understandings into expanded systemic ones. Alshuler noted that the process of withdrawing projections is a major developmental task within both these developmental models because it facilitates a fuller understanding and acceptance of ourselves and others in ways that are ultimately empowering. Embracing this maturational task in order to constructively work with climate issues, was an essential part of the group's discussions about what it means to live responsibly, ethically and well within a climate disrupted world.

Developing an enlarged awareness that supports holistic understandings through an ability to observe projective dynamics at play in ourselves and others, does not mean absolving government or corporate powers from their responsibility for climate disruption. It rather builds foundations to consciously work with the interconnectedness between individuals and systems in order to address the complex and messy business of our mutual destructions. In recognising that our ways of living were both part of the problem and the solution, participants encouraged one another to relinquish simplistic positions in order to front up to the kinds of tensions and choices that we encountered daily. Lisa observed how participation in the group helped her be active in the world in a different way by learning to hold her space at work, "I've learnt not to be judgemental about it – to let go of that hideous anxiety that I have been feeling that it's deliberate somehow. It's not! It's just... a different worldview". At the same time, acknowledging the limitations of personal agency within a systemic problem of complex dimensions helped to inform participants' commitments and actions for sociopolitical change. For Zoe, this meant becoming "much more concrete around my response and my direct action [because] I have greater trust in how I make decisions, and how my behaviour affects bigger questions".

While psychological and socio-political perspectives cannot be collapsed into another, they can inform and enrich each other, when not approached as binaries (Alshuler 2006; Samuels 2001). In this research, the psychological focus of its discussions stimulated a personal maturational development (individuation) which often also supported the development of informed political analysis (conscientisation). At the same time, analyses of the socio-political dimensions of climate issues, as understood by participants, helped to interrogate personal relationship to the world, extending self-knowledge and emotional capabilities. Simon observed at the final meeting that:

preparedness for change isn't solely about the actions you do, but there is this important psychic level, and that what we were doing was very worthwhile on that level too… that was a really useful and satisfying way of looking at the group in terms of that kind of preparation, and potential connection between personal attitudes and potential political stuff.

Through staying present to both personal and public dimensions of climate disruption, the group process helped to dismantle artificial constructions which isolate and diminish individual identity and agency, bringing consciousness to the collective dimensions of life.

By accepting the realities of individual limits and smallness, at the same time as exploring understandings of, and connections to, systemic and global perspectives, participants extended their perceptions beyond singular beliefs and viewpoints, feelings of isolation and divisive positions. As a result, a greater relatedness became possible, internally and externally, building emotional intelligence, resilience and common ground for response and action.

Conclusion

Naomi Klein (2014) entitled her groundbreaking book on climate change *This Changes Everything*. Her title is equally apt for the psychological terrain of individuals and societies as it is for the political economies of the world. Just as Klein perceives the potential for climate disruption to act as a catalyst for positive global socio-political change, so too is there potential for it to spark positive psychological transformations in individuals and communities. In order for this to happen however, there needs to be the stimuli and spaces for open dialogues, both formal and informal, which encompass and proceed beyond the old binaries of inner/personal/individual life and outer/social/collective life.

The collaborative research presented here demonstrates that ongoing reflective conversations attentive to the psychological dimensions of climate engagement can stimulate a development of consciousness that increases maturity and resilience. Through exploratory discussions, participants were able to acknowledge contradictory desires and conflicts, re-examine core values, confront existential anxieties and work through the challenging emotions of profound change. Out of this process, new perspectives evolved which supported flexible, innovative and empathic responses to the dynamic and complex ecological, socio-political and psychological tensions of climate disruption.

The widespread failure of governments to address climate disruption is stimulating grass root initiatives and forums. Each provides opportunities for conversations which nurture consciousness change and increased empathy (see the Hamilton chapter in this volume). Climate discussions that transform people's hearts and minds need to be personal with shared feelings, stories and imaginings. They invite connection and inclusiveness while opening up existential questions of value and meaning in life. Respectful and exploratory dialogues allow deep-seated assumptions to be examined in a non-threatening way and for the commonality of where we are and who we are to emerge. Bringing consciousness to conflicts and connections, within ourselves and between ourselves, lays foundations for collective action based upon a diversity of perspectives, a network of relationships and an ethics of care for our world.

References

Alshuler, L. R. (2006). *The psychopolitics of liberation: Political consciousness from a Jungian perspective*. New York, NY: Palgrave Macmillan.

Bosnak, R. (2007). *Embodiment: Creative imagination in medicine, art and travel*. New York, NY: Routledge.

Boyd, R. D. (1991). *Personal transformations in small groups*. London: Routledge.

Bulkeley, K. (2008). *American dreamers: What dreams tell us about the political psychology of conservatives, liberals and everyone else*. Boston, MA: Beacon Press.

Connor, L. H. (2011). Anthropogenic climate change and cultural crisis: An anthropological perspective. *Journal of Australian Political Economy, 66*, 247–267.

Coppin, J., & Nelson, E. (2005). *The art of inquiry: A depth psychological perspective* (2nd ed.). Putnam, CT: Spring Publications.

Fisher, A. (2013). *Radical ecopsychology: Psychology in the service of life* (2nd ed.). Albany, NY: SUNY Press.

Gillespie, S. (2013). Climate change and psyche: Conversations with and through dreams. *International Journal of Multiple Research Approaches, 7*(3), 343–354.

Heron, J. (1996). *Co-operative inquiry: Research into the human condition.* London: Sage.

Heron, J., & Reason, P. (2008). Extending epistemology with co-operative inquiry. In P. Reason & H. Bradbury (Ed.), *Sage handbook of action research: Participative inquiry and practice* (pp. 366–380). London: Sage.

Hoggett, P., & Randall, R. (2016). *Sustainable activism: Managing hope and despair in social movements.* Open Democracy. https://www.opendemocracy. net/transformation/paul-hoggett-rosemary-randall/sustainable-activism-managing-hope-and-despair-in-socia.

Hillman, J. (1992). *Revisioning psychology.* New York, NY: HarperPerennial.

Jung, C. G. (1969). *The structure and dynamics of the psyche* (2nd ed., Vol. 8). London: Routledge, Kegan and Paul.

Jung, C. G. (1974). *Dreams.* Princeton, NJ: Princeton University Press.

Klein, N. (2014). *This changes everything: Capitalism vs the climate.* New York, NY: Simon & Schuster.

Lifton, R. J. (1996). Dreaming well: On death and history. In D. Barrett (Ed.), *Trauma and dreams* (pp. 125–139). Cambridge, MA: Harvard University Press.

Madriz, E. (2003). Focus groups in feminist research. In N. K. Denzin & Y. S. Lincoln (Eds.), *Collecting and interpreting qualitative materials* (2nd ed., pp. 363–388). Thousand Oaks, CA: Sage.

Paris, G. (2007). *Wisdom of the psyche: Depth psychology after neuroscience.* East Sussex, NY: Routledge.

Randall, R. (2012). Fragile identities and consumption: The use of "Carbon Conversations" in changing people's relationship to "stuff". In M.-J. Rust & N. Totton (Eds.), *Vital Signs: Psychological responses to ecological crisis* (pp. 225–238). London: Karnac.

Reason, P., & Bradbury, H. (2001). Inquiry and participation in search of a world worthy of human aspiration. In P. Reason & H. Bradbury (Eds.), *Handbook of action research: Participative inquiry and practice.* Thousand Oaks, CA: Sage.

Samuels, A. (2001). *Politics on the couch: Citizenship and the internal life.* London: Karnac.

Samuels, A., Shorter, B., & Plaut, F. (1986). *A critical dictionary of Jungian analysis*. London: Routledge, Kegan & Paul.

Watkins, M. (2008). "Breaking the vessels": Archetypal psychology and the restoration of culture, community and ecology. In S. Marlan (Ed.), *Archetypal psychologies: Reflections in honour of James Hillman* (pp. 415–437). New Orleans, LA: Spring Journal Books.

Watkins, M., & Shulman, H. (2008). *Towards psychologies of liberation*. Basingstoke: Palgrave Macmillan.

7

Climate Change, Social Dreaming and Art: Thinking the Unthinkable

Julian Manley and Wendy Hollway

Beyond the Scientific Fact

The problem of how people can accept the reality of climate change and its effects on our daily lives is central in climate psychology. Scientific facts have proved remarkably ineffective in leading to necessary changes in lifestyle required on both an individual and a social level. For many, the facts are either traumatic or unacceptable. The requirement posed by global warming to change people's deeply held desires for ever-increasing economic prosperity and the assumed concomitant wellbeing leads to shared and generalised disavowal and denial. In the world of climate change deniers or disavowers the status of scientific factual reality is a significant issue: the scientific facts backed by 97% of the

J. Manley (✉)
University of Central Lancashire, Edinburgh, UK
e-mail: jymanley@uclan.ac.uk

W. Hollway
Open University, Hebden Bridge, UK

© The Author(s) 2019
P. Hoggett (ed.), *Climate Psychology*, Studies in the Psychosocial,
https://doi.org/10.1007/978-3-030-11741-2_7

scientific community are not 'fact-enough' for meaningful social change: information, debates, surveys, focus groups and suchlike fail to open the way to significant action. In the case of climate change we are in a zone of gut rejection: even if it is, it cannot be. Al Gore's 'inconvenience' (Gore 2006) is more than that: it is something so inconvenient that it cannot be countenanced.

Psycho-social approaches to climate change, therefore, tend to take a containing approach to people's fears, traumas and deep concerns. For example, Randall and Brown's (2015) 'carbon conversations' project provides practical and experiential psycho-social approaches designed to create contained spaces for reflection and transformation. Through conversation, according to Westcott (2016), there is a chance for denial and disavowal to be converted into hope and trust, without which climate anxieties are repressed and ignored rather than confronted. Such approaches have been positively evaluated by Buchs et al. (2015) who summarise the emotions that can be discussed through conversation related to climate change as fear and anxiety, grief, guilt, helplessness and feeling threatened in one's identity/status (Buchs et al. 2015, p. 622). It is through the careful containment of shared conversations that people are given an opportunity to be released from the isolation, loneliness, guilt and even horror that scientific facts point to. These conversations change the nature and quality of the climate fact through each person's *relation to* the facts. In a sense, the reality of the fact is given a potential for being re-experienced, almost as if it were not a factual entity in and of itself. Climate facts are thus subjectivised and their reality is found in the transactions between external and internal world experiencing.

This chapter concentrates on a different way of knowing, focusing on the shared visual and affective aspects of people's relationship to climate change. It uses the data from an art and social dreaming event to explore how the use of affect-laden images in a shared 'unconscious' context, hidden or unknown, can help us to recognise the reality of climate change. Social dreaming is a method that allows new knowledge to emerge in a gathering of people who share their dreams, associations and feelings together. The method creates a non-threatening, non-judgmental space where difficult thoughts and feelings can be expressed through images (Lawrence 2005; Manley 2014, 2018). In Social dreaming and the visual arts, the realm of worded communication is subsumed into a world of image and affect. Both involve what Donald Meltzer calls the 'poetry of the dream' whose role in thinking is that it 'catches and gives formal representation to the passions which *are* the meaning of our experience so that they may be operated upon by reason' (Meltzer 2009, p. 47).

According to Meltzer (2009 [1984]) 'Dream life can be viewed as a place to which we can go in our sleep, when we can turn our attention fully to this internal world'. He uses the phrase 'dream life' to emphasise the kind of dreaming that goes on all the time, awake or asleep. The idea that dreams are part of our everyday existence was also an important aspect of Bion's concept of the unconscious (Ferrero 2002). The social dreaming matrix in its function as a container of dreams and associations in this way provides a space that enables more than a mere recounting and description of past dreams. It is, rather, a reliving of dreams, what Ferrero calls 'redreaming', a 'syncretic narrative mosaic' (Ferrero 2002, p. 605) that brings the dreams back into present experience, thus creating a 'dream life'. It is to these internal worlds that the Social Dreaming method listens, using the waking recounting of night dreams and associations to them, supported by facilitating reverie in a containing space. 'The creative process of dreaming generates meaning that can then be deployed to life and relationship in the outside world' (Meltzer 2009 [1984], p. 46). Meltzer is explicit about the central place of the aesthetic in psychoanalysis, describing his account of dream life as an attempt to formulate an aesthetic theory of dreams (Harris Williams 2010, p. 8). In our social dreaming/art event, the experience of engaging with the artworks was combined with the dreams in the matrix, engaging the participants in a presently shared lived aesthetic of dreams and associations that opened up a new space of creative thinking leading to new meanings. The way shared dreams can naturally combine with shared experiences of art resides in how both experiences are able to provide an affective filter for past experiences and current knowledge to be evoked, shared, reconsidered and reconfigured as new knowledge *which is no longer past.* The dreams are processed internally and shared externally—inside out—while the artworks are received externally and processed internally—outside in—and join the dreams in an eventual shared experience in the matrix. This inside-out/outside-in flow of image and affects of the aesthetic experience can be compared to Guattari's identification of such a psycho-aesthetic process as '*processual complexity*, with possibilities that are constantly developing' where the emphasis is on the present, 'not on the past' (Guattari, quoted in Ettinger 2002, p. 243, author's italics).

The artworks experienced by participants in this social dreaming event were curated by the arts organisation, Cape Farewell (www.cape-farewell.com). Their mission—to mobilise climate change awareness through the medium of art—is based on the same belief that art is capable of effecting change by engaging people in the kind of emotional responses that fact-based interventions fail to do. This belief is difficult to establish systematically through research, a key problem being that dominant research paradigms are science-based, hence impose a non-congruent paradigm in relation to the humanities, art and culture. Epistemologically this problem is often conceptualised as the sciences and arts drawing on different types of knowing: rational/emotional, factual/aesthetic, particular/holistic, logical/intuitive, explicit/implicit. Here we explore a paradigm that is congruent with art, culture and the humanities.

Social Dreaming and Climate Change

The principles we are adopting here to inform Social Dreaming, contrary to the compartmentalisation that characterises scientific methodology, have the capacity to explore an indivisible experience; a different kind of knowing. In different psychoanalytically informed paradigms, these principles are referred to as affective, associative, rhizomatic or scenic. Psychoanalysis has provided an ontology that is comfortable with the non-cognitive realm, with dynamic, unconscious and conflictual forces. It understands aesthetic experience as outside the cognitively aware realm. Ehrenzweig (1967) theorised the apprehension of art as taking place at two levels, which he termed analytic and syncretistic, the latter being the indivisible experience. Social dreaming has these qualities but goes further by not being limited to the individual focus that has characterised much of psychoanalysis historically. It generates knowledges that go beyond the individuals who feature in the research design and constructs shared imagistic social collages of experiences that cannot adequately be put into words. The dreams, their associations and the affects embedded within them, become recognisable in social dreaming as images in a spontaneous, nomadic landscape, not as lines of cognition in words, thoughts and feelings. 'Nomadic' here refers to the Deleuzian account of concepts and creations that emerge without restriction or preconceived directions, ideas in constant travel and in ever-changing states of emergence (Deleuze and Guattari 1988). These reach into places of inarticulable knowledge—the realm of the 'unthought known' (Bollas 1987)—based on direct experience, that is poorly represented by consciously available climate change discourses.

From a largely humanities base, the idea of climate change as a 'hyperobject' emphasises this impossible-to-grasp nature of climate change. Hyperobjects—of which global warming is Morton's (2010, 2013) leading example—have an extension in time and space that makes them historically beyond the range of human cognition; they 'massively outscale us' (p. 12). However, although the idea of global warming in its entirety cannot be accessed at a cognitive distance, it is 'right here in my social and experiential space' (2013, p. 27), having a viscous quality: climate change 'sticks' to everything—the plastic food wrapping, the car journeys, heating the house, the microwave, the smell of bacon, waste disposal... Paradoxically, then, the hyperobject is also non-local because its local direct manifestations are not the hyperobject itself (although it is raining heavily again this is not 'global warming' that I feel directly). These characteristics require a different research approach than science and social science afford. A humanities-based research paradigm should produce different forms of knowledge. If our methods can access the 'unthought known' and the hyperobject of climate change, what does this look like and how can it be made accessible to wider culture?

Design

The design of the event was innovative It incorporated artworks into the dreaming experience: An hour in the evening was spent viewing specially curated[1] artworks, which made some reference to the effects of climate change. They included visual images, objects, texts, filmed performance poetry, sculpture and pottery. This research design enabled us to look at the part played by the artworks in the dreams and dream associations that emerged in the Matrix. The participants were climate activists—artists, scientists, researchers, and practitioners—people who face the threat of climate change on a continuous basis.

Context and Process

On 13 October 2017 sixteen participants gathered at a farm in Dorset, where in the surrounding barns there were climate-focused artworks from the Cape Farewell archive. On the first evening, the purpose of the event was outlined and participants were invited to spend an hour sharing and engaging with each other in the context of the farm buildings, where new social and aesthetic spaces were created, 'transitional' spaces in the Winnicottian sense (Winnicott 1991). Next morning, the Social Dreaming Matrix was convened and one of the two 'hosts' (the common term for facilitator in social dreaming events)—opened with the standard question 'what is the first dream?' This question deliberately emphasises the dream, not the dreamer, reflecting the fundamental claim of the method that its elicitation of associative thinking goes beyond individual experience and meaning to produce a wider, multi-layered knowledge that we might call socio-cultural. This approach helps us to reject the compartmentalisation, reduction and consequent stripping of affective meaning typical of the scientific approach, which increasingly is at odds with the dawning of the age of

---

[1]Curated by David Buckland on behalf of Cape Farewell.

the Anthropocene, and fashion something more in line with the Gaian principle of indivisibility (Lovelock and Margulis 1974).

During the process of the social dreaming matrix, conditions for reverie (the waking dreaming state that affords associative thinking) are carefully created: low lighting, quiet and a 'snowflake' seating pattern that discourages face-to-face engagement. These features help associations, images and ideas to emerge in place of facts, opinions, argument and cognition-based interactions. The matrix is followed by a facilitated post-matrix discussion that provides a space for a more cognitive mode of reflection. Both are recorded, and the ensuing transcription is used in analysis by the researchers.

Sequences of Association

To provide familiarity with the images and associated ideas that make up the matrix collage, we present an abbreviated chronological treatment of the first half of the matrix which, rather than being simply descriptive, traces a sequence[2] that glimpses the passage of ideas from a modernist machine age of fossil fuel consumption through turbulent and destructive change to an imagined post-anthropocentric regeneration of the earth. This 'historical' sequence is salient in Couze Venn's (2018) *After Capital*, which situates climate change as central to converging world crises of capital, manifest in 'the global economy, the environment, the depletion of non-renewable resources, increasing violences and the break-down of social cohesion'. These 'threaten a perfect storm in the near future' (Venn 2018, p. 1). Given that all the participants are living through these changes; given that there is not something sealed off called 'global warming' or 'climate change', how did this show up in the matrix—if it did at all? The social dreaming method

---

[2]Although the idea of a linear sequence is antithetical to our conceptual framework, there is a rationale for proceeding in this way, namely that the matrix proceeds by associations. The sequence is not always linear, as we discuss, but there is a sequence, which necessarily begins in linearity at the start of the matrix. At a later stage in the matrix that linearity gives way to rhizomatic structures.

claims to capture something socio-cultural. The following analysis suggests that it did.

Gordon Lawrence, the inventor of Social Dreaming, believed that the first dream presented in a matrix was important in that it acted as a thematic forerunner to the whole collage sequence (Lawrence 2005, p. 15). The first dream went as follows:

> I remember hearing that a friend's Great Aunt Vera was driving and that she shouldn't because she is so frail and I had to do something about it, but I didn't know where she lived, apart from the fact that it was in Devon. So I went to a local post office store and asked the lady there if she could help. She said she was bound to know someone who could track her down. She went into the back office in the back of the shop and came out again and said "Sorry, we haven't been able to find Vera, but I believe she isn't driving, but rowing down a river". I said "Vera? Rowing down a river? She's far too frail for that!" Apparently the way she had done it was she had an inflatable snowman in the bow of the boat with two ropes attached to it, and if she pulled the snowman forward and let it go, it sprung back and rowed the boat... end of dream.

Let us treat Great Aunt Vera and her activities as a metaphor for humanity on a frail planet: the name signifies wisdom, age, experience and feminine gender—truth (Vera), generation (Aunt) and importance (Great). Driving references the machine age and fossil fuel consumption, which of course is alarming for one so frail. The dreamer 'had to do something' to stop this driving. Devon references a part of the UK where nature is unspoilt and local populations value it; the post office lady is part of a community where people still communicate, support each other. Perhaps the back office is where a different kind of knowing can be accessed, in contrast to the monetised transactions of the front counter. The information is only partially successful: Vera is not within their sights (perhaps this refers to how hard it is to pin down global climate change) but local knowledge has come good: she/the wisdom of old humanity is transformed, discarding machine age transport and accomplishing something more suited to youth than age, a result of ingenuity and natural energy. There is also an implication that a natural force can 'bounce back' when captured by humans (ropes) or that the planet can bounce back once carbon is under control. Great Aunt Vera is transported by ecological methods, borne along on water rather than tarmac. The snowman references the snow that is perishing at the earth's poles; inflatable (what is inflated? Claims of snow melt?) and vulnerable.

So the first dream reveals, in fractal fashion, a climate change scene: planetary frailty and wisdom, natural resources and technical ingenuity, obsolescence and innovation, human responsibility, community and concern, not yet tracked down but knowledge from the 'back'.

Water was a recurrent theme throughout the matrix; the two dreams following 'Great Aunt Vera' involved water borne transport—windsurf and yacht. In both dreams a storm/rough sea/big wind figured. In the second of these the defunct status of machine transport figured in 'hulks of buried helicopters' below the sea. On one hand, a high wind meant that 'the boat is sailing along at incredible speed' but there is ambivalence 'I even had the engines on as well in order to increase the speed, but then thought "that's stupid" and switched the engines off, the helicopters were still there'. Just because fossil fuel based transport is no longer necessary, doesn't mean to say that people aren't still wanting to use it—for extra and unnecessary speed. Soon followed dreams of 'a crappy one carriage train' that had already left the station and 'a cartoon airplane face up in the water'; then immediately another image of defunct machine age transport: a Boeing 747 with no wings or tail, parked on the road, taking up a lot of room but unnoticed by people, a detail that references how taken for granted fossil fuel based transport is even when it is blatantly dysfunctional and wrongly situated.

Segued from the theme of transport came a dream that got to the heart of climate change threat: travelling by plane and car along hot and dusty routes, to the funeral of Mother Earth. But not empty handed: carrying a large brown nut, which was meant to be delivered there. It was in a white bag with 'palms' printed on the side. The nut is a seed, symbol of regeneration and the capacity of nature to carry on. Likewise a golden haired boy keeps turning up, for example at the airport; a symbol of the responsibility of an older generation to keep faith and deliver the hope of the earth's survival, a symbol that recurred many times, expressed as young women, young employees, a baby in a pushchair, a group of students and a toddler.

The theme of danger and disaster was continued in further dreams with Alpine avalanches killing unconcerned tourists in Wales and a nurse offering an inadequate remedy to a man 'completely full of water'. There followed an association to 'water water everywhere' (usually connoting 'not a drop to drink'). The water images multiply in the condensed fashion of dream images: flooding, sea level rise and drinking water scarcity resulting from drought. Eventually we are led to an image of water cleansing a turning globe, getting rid of 'us', the human beings responsible for the damage.

Expansion

At this stage of the social dreaming matrix, nearly half an hour in, the gentle chronological linearity in the matrix—which mirrors the sense of history that brought us from the taken-for-granted progress of the fossil fuel era to the current era of negative consequences—is challenged with a dream where the dreamer hits a signless sign post 'right between the eyes'.

> I was walking and a lamp post post suddenly hit me between my eyes. […] It wasn't a lamp post actually, it was more like a sign post, just a post without a sign on it.

This brief dream is an ironic wake-up call which plunges the matrix into deeper reverie in order to 'find the way' symbolised by the signless sign post. As a transition into this new direction, there is a realisation within the matrix that the issue of climate change has gone as far as it can in the old direction. A hit needs to be understood somehow through the body—'right between the eyes' (the 'third eye'?), which connotes sudden insight. A series of four references to the body immediately ensues:

> (1) I had the power to rearrange the body... (2) This is bringing up this line from Boris Johnson he said about clearing away the bodies... (3)

My body dictates the texture and quality of my dreams… (4) If you can't
sleep … you pour cold water on your feet and legs, just that part of your
body…

The matrix then launches into a rhizomatic series of associations that
combine the artworks and the dreams and associations, thus breaking up the general linearity of the matrix. A renewed emphasis on the
body reassesses the role of affect, that is to say knowledge that is felt
by the body before it is thought. The matrix expands into creativity, a
new energy, by offering an alternative to fossil progress, which, as represented in the early references to machine transport, is the cause of climate change. The alternative was hinted at in the opening dream magic
of Great Aunt Vera's boat, which is referred to once again in this new
rhizomatic section of the matrix:

Energy repelling energy reminds me of Aunt Vera's snowman and the
peculiar, magical form of motion.

The way the matrix is enabled to discuss alternative ways of perceiving
and feeling human knowledge, described here as 'energy', is precisely
through a combination of dreams, artworks, associations and affect,
which we analyse below.

Data Analysis and the Rhizome

Data analytic principles need as much modification to be suitable to
our purpose as did the approach for generating the data: not linear,
not broken up into discreet entities, respecting the bizarre-seeming,
the multiple and uncertain, sensing the affect, following the associative
unconscious through links and gaps. This 'rhizomatic' analysis is guided
by Deleuze's image of a 'rhizome' as a randomly self-selecting set of
interconnected image-affects that pulse in intensity at a given moment
of perception. Instead of lines of thought, Deleuze and Guattari posited
the concept of the rhizome to describe a 'circulation of states' (1988,
p. 21) that have no predestined direction, no centre, no hierarchy or

guidance from any authority. This description reflects the structure of the collage of associations of the social dreaming matrix, its fluidity and the way it is created, namely through spontaneous expression of dreams, affect and associations of participants.

In this view, the chronological 'beginning' is not necessarily the best place to start the analysis, although our method is limited by the irreducible linearity of language that forms our transcript, which must be read in the context of time passing. We have shown above how a certain sequentiality of associations led on from the opening dream and how this can be justified both in terms of the nature of the interactions and the chrono-historical story of the journey of fossil-led progress. In what follows, we delve into the body of the matrix and show how linearity gives way to rhizomatic configurations.

At this stage of the matrix, the contributions seem to pass a syncretic threshold of expression: the dream images and associations begin to fuse and merge with the artworks. At the same time, the dream images from the beginning of the matrix also re-emerge at this point with a different and greater intensity than when they were first presented to the matrix. It is at this stage that the complexity of the affective interchanges of the matrix appear to mirror the inherent complexity of climate change as a hyperobject, where, in rhizomatic fashion, the earlier experience of witnessing the artworks is ignited in the thoughts and feelings of the here and now of the social dreaming matrix. As this merger occurs, there is a sense that the artworks, as they exist in the minds of the participants in the matrix, become available to participants in much the same way as the dream images are functioning for and in the matrix, thus seemingly acquiring a dreamlike quality and becoming a seamless part of the increasingly complex collages of images that form the thinking and affective space of the matrix.

As the matrix proceeds knowledge emerges in an implicit manner reminiscent of Polanyi's 'tacit knowledge' (1967) or Bollas' 'unthought known' (1987) or Bohm's 'implicate order' (1980). These theoretically diverse terms all indicate a similar preoccupation with what is hidden from our cognitive function, what in psycho-social studies is sometimes referred to as knowledge that lies 'beneath the surface' (Clarke and Hoggett 2009), from some less-than-conscious space in the shared arena of affect that is the social dreaming matrix. These thoughts and affects are extremely complex as they emerge from that implicate, unthought place, where the multiple reality of complex systems—the hyperobject of climate change—can be made known through the sharing of hinted at and beneath-the-surface knowledge. It is not the finer detail that drives the message home, rather it is the accepted wholeness of an intuited sensation of resonating congruity and affect shared among the temporary friends who form the matrix, where awareness of the rhizome of affects does not demand or presuppose a cognition of its every node and movement.

A Pulse of Intensity and the Creativity of the Rhizome

At the half hour point of our social dreaming matrix there is a significant shift in the development of the rhizome of affect that is being shared as the dreams and artworks unexpectedly converge. The artwork that emerges at this moment in the matrix is the following by David Buckland, which juxtaposes a hugely magnified image of a coccolithophore (a single cell phytoplankton whose calcium carbonate structure is the substance of all chalk cliffs), with a scientific description by the scientist Iglesias-Rodriguez of a carbon-based process in seawater that contributes to global warming (Fig. 7.1).

The beautiful image of the coccolithophore that is embedded in the artwork is under normal circumstances invisible to the naked eye: it cannot naturally be seen with the eyes in our heads. We might know it is there (and this artwork provides us with such knowledge), or we might intuit it, and this reminds us of the signless sign post, where not everything can be seen and therefore it is unclear what direction one should take. To 'see' what is hidden in this requires the artwork. To make sense of it and discover a 'new direction' requires the fusion of the artwork into the collage of images of the matrix.

In their inscrutability, the *words* provide a false sense of direction. The *image* that cannot be seen is nevertheless revealed through magnification

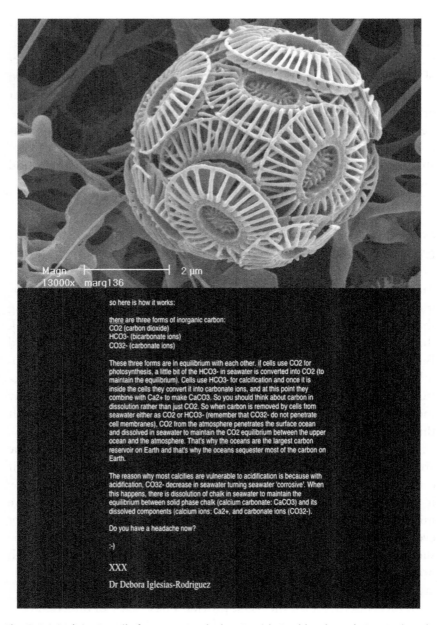

Fig. 7.1 Caché. Detail from artwork by David Buckland and Dr. Deborah Iglesias-Rodriguez. www.bucklandart.com/art/cache-2012/

and displayed as an artistic image rather than an illustration of science; the scientific text, presumably logical, rational and explanatory, is also displayed as art. It can be 'seen' and yet remains indecipherable to the non-scientist despite the aim of science to verbalise with absolute clarity. In different ways, neither image nor text can be 'seen'. Ironically, and to make the point, the text is completed by a winking emoji (itself an image directly linked to an emotion, rather than a word) and the line 'Do you have a headache now?' This speaks to Morton's idea that 'the more data we have about hyperobjects, the less we know about them' (2013, p. 180). Conversely it might also suggest the possibility that the fewer facts at your disposal, the more you know. This is the case of the image in the artwork. The fact of the science is replaced by an aesthetic experience. It is through the image as an artwork rather than as fact that participants come to 'know'.

The turning point that connects the dreams to the artwork is when one of the participants recounts a dream where he is hit 'right between the eyes', mentioned above. From this moment the dreams and artworks are fused within the body of the matrix. In the dream offered at this stage, an adult is climbing up the stairs of a tall tower-like building, 'entrusted with the care of some toddler'. They both manage to climb the insecure stairs despite the dangers of a lack of bannister and gaps that must be bridged with ladders. All of this is achieved with 'quite a feeling of omnipotence'. In this dream, the success of the climb and the handing down of experience and knowledge through the generations as represented in the 'toddler' elicits a sense of superior human conquest in overcoming obstacles. It is the ultimate example of humankind's rise (up the stairs) and unparalleled success in defying the odds (the gaps) along the path of progress that the matrix has been concerned with until this moment. The dream is immediately followed by the following association to the artwork:

> The image of the stairs reminds me of the image of the carbon molecule (sic) on the barn wall outside, incredibly intricate and beautiful structure, which I found very moving, and then I noticed one of the ladders on the stairs was broken or only half formed in this microcosmic molecule, and I was wondering what kind of stresses that expresses or whether they

were never perfectly formed for some reason. But that image of the broken stairs reminded me of the imperfect but perfectly beautiful cellular structure.

In this association that is explicitly connected to the previous dream, artwork and dreamwork come together to uncover the hidden and mysterious beauty of the 'molecule' that can potentially be found in the dream structure of broken stairs winding their way upwards. Where in the dream we are left with a sense of conquest (omnipotence), now our attention is drawn to something 'incredibly intricate' and 'beautiful'. Importantly, this perception of the hidden (made visible through magnification), is described as 'very moving', an aesthetic experience. So here we have a description of felt affect and its embeddedness in the visual that is fundamental to the Deleuzian rhizome. Instead of broken stairs being gaps to overcome as in the dream, the 'broken or only half formed' stairs in the perception of the image in the artwork are inducements to affective reflection, a questioning of the kinds of 'stresses' that might have brought these about. The joining of the associative affect of this reflection on the artwork with the telling of the dream brings into question the conquering tone of the dream noted above, and presents a possibility of some other direction to take: maybe inwards (as in a magnification) rather than outwards/upwards, as in the dream. It provides a new language and sign for our signpost.

As a result of this new direction, the matrix can move on to a different, creative and affective pursuit of knowledge. This knowledge is summarised in the following reference to 'energy' as part of the new development of the associations provoked by the fusion of artwork and dreamwork:

> That reminds me of images of molecules and that most of the molecule or atom is space, there's nothing in it except energy.

This 'energy' appears to suggest an invisible (now acceptable in the thinking of the matrix) binding agent between disparate elements that is reminiscent of the Deleuzian concept of affect as a continuum between objects that increases/decreases in intensity and that holds elements together in relationship according to this continuously fluctuating pulse of affect. The energy evoked in the matrix and stimulated by the thinking and feeling process of the dreams and the experience of the artworks can be thought of in terms of Deleuzian 'duration' that operates between states of affect in rhizomatic relationship. These states are, as Deleuze comments with reference to Spinoza, 'affections, images or ideas [which] are not separable from the duration that attaches them to the preceding state and makes them tend towards the next state' (Deleuze 1988, p. 49). The establishment of the rhizome that constitutes the structure of this non-linear interconnectivity between the dream images that now merge with the artworks is 'seen' in the embodied mind of each participant. This effect is exemplified in the following contribution that makes an association with the first dream of the matrix; in other words the structural pattern in the mind has become a rhizomatic connection rather than a sequential one, linked by affect or 'energy':

> Energy repelling energy reminds me of Aunt Vera's snowman and the peculiar, magical form of motion.

Once established in this new mode of reflection, the matrix is able to see the combination of affective and associative thinking as more than theoretical and as actually defining the embodied process of the matrix itself in the following reference to the snowflake configuration of the here and now of the matrix:

> It's funny because it looks really close but also a feeling of space like the atom you were talking about: the way of talking and sharing in the now.

A further association describes this space of new affect which is contained in the created rhizome of the matrix as being a safe space for thought and, crucially, a space where the 'huge' can be contemplated:

> A lot of dreams have involved catastrophes and disasters, and things going wrong, but always as us the observers in a safe place, which is interesting

because that sort of feels like what we're doing here, we're talking about or thinking about a huge issue but feeling fairly safe.

The ability to talk of huge and disturbing issues in sufficient safety suggests that social dreaming and art did indeed create a space of understanding of the Mortonian hyperobject.

Conclusion

At the start of this chapter we set out to discover whether the combination of social dreaming and the experience of artworks related to a theme of climate change could provide a means of capturing a shared unthought known of climate change, capturing its enormity and complexity in such a way as to make the possibility of effective and meaningful action a genuine possibility. We posited that scientific facts were not effective for change, as demonstrated in the failure to elicit lifestyle and social changes that would help to stop and reverse climate change, despite overwhelming scientific factual evidence that should leave no room for doubt. We suggested that we needed a new epistemology, a new way of expressing and perceiving knowledge, and our study has pursued the idea that such a knowledge may be brought into shared consciousness through social dreaming and art.

In investigating social dreaming and art as a means of reconfiguring knowledge, our material has shown the utility of re-imagining a new sense of 'fact', moving from a bare, reduced raw fact of science to what is meaningful to participants in the matrix. The latter is like a felt truth, where the eventual knowledge is sensed through affect and made into a different 'fact' through the sharing and relatedness of the participants' experiences of the artworks and social dreaming. The nature of the hyperobject is not that it is free of facts but on the contrary it embraces too many compartmentalised facts to cognitively capture, understand and eventually to act upon. It is truly complex. It cannot be seen. In this way we unpack Morton's philosophical take on the difficulty and yet the need to understand climate change as a hyperobject. The hyperobject, we have pointed out, is felt in aspects that are associated and yet not directly connected to the scientific fact of climate change. These associations of the social dreaming matrix are no less true than the facts of the hyperobject but they present these truths in a way that can be seen and felt. An example of this in the matrix was the creation of a 'fact' of the 'energy' that holds ecological complexity together. The identification of this energy made it possible for participants to understand how each separate aspect related to climate change can assume a connected unfragmented whole. It made the interconnectivity of Gaia theory more tangible. An important pathway to this grasping of the hyperobject was in the act of social dreaming itself, which, we have suggested, is a practical method of 're-dreaming' night time dreams and connecting to 'waking dreams' that are better able to give meaning to the hyper and the macro. In doing so, we have drawn on Meltzer's and Bion's theories of the dream world as being omnipresent, as relevant to waking life as sleep. Meltzer and Deleuze and Guattari have linked the creative imagination to the aesthetic experience. We saw how the aesthetic experience enabled the social dreaming matrix to become complex in rhizomatic intensity mirroring the complexity of climate change and providing the conceptual and experiential shift in knowledge from linearity to complex systems that can be experienced in the here and now of the social dreaming matrix. Importantly, therefore, we conclude that it is through this associative thinking in a shared context that new ways of knowing are created, an experience of felt knowledge nurtured into existence through the combination of social dreaming and art, which enables us to apprehend climate change in the macro-cosmic way that it demands and gestures to an imagined and almost hopeful post-anthropocentric vision.

References

Bohm, D. (1980). *Wholeness and the implicate order*. London: Routledge.

Bollas, C. (1987). *The shadow of the object: Psychoanalysis of the unthought known*. London: Free Association Books.

Buchs, M., Hinton, E., & Smith, G. (2015). "It helped me sort of face the end of the world": The role of emotions for third sector climate change initiatives. *Environmental Values, 24*(5), 621–640.

Clarke, S., & Hoggett, P. (Eds.). (2009). *Researching beneath the surface.* London: Karnac.

Deleuze, G. (1988). *Spinoza, practical philosophy.* San Francisco: City Lights.

Deleuze, G., & Guattari, F. (1988). *A thousand plateaus.* London: Continuum.

Ehrenzweig, A. (1967 [1970]). *The hidden order of art.* London: Paladin.

Ettinger, B. L. (2002). From transference to the aesthetic paradigm: A conversation with Félix Guattari. In B. Massumi (Ed.), *A shock to thought* (pp. 240–246). London: Routledge.

Ferrero, A. (2002). Some implications of Bion's thought: The waking dream and narrative derivatives. *The International Journal of Psychoanalysis, 83*(3), 597–607.

Gore, A. (2006). *An inconvenient truth, film.* Beverly Hills, CA: Lawrence Bender Productions.

Harris Williams, M. (2010). *Bion's dream: A reading of the autobiographies.* London: Karnac.

Lawrence, W. G. (2005). *Introduction to social dreaming: Transforming thinking.* London: Karnac.

Lovelock, J., & Margulis, L. (1974). Atmospheric homeostasis by and for the biosphere: The Gaia hypothesis. *Tellus, XXVI.* http://www.jameslovelock. org/atmospheric-homeostasis-by-and-for-the-biosphere-the-gaia-hypothesis/.

Manley, J. (2014). Gordon Lawrence's social dreaming matrix: Background, origins, history and developments. *Organisational and Social Dynamics, 14*(2), 322–341.

Manley, J. (2018). *Social dreaming, associative thinking and intensities of affect.* London: Palgrave Macmillan.

Meltzer, D. (2009). *Dream life.* London: Karnac.

Morton, T. (2010). *The ecological thought.* Cambridge, MA: Harvard University Press.

Morton, T. (2013). *Hyperobjects: Philosophy and ecology after the end of the world.* Minneapolis: University of Minnesota Press.

Polanyi, M. (1967). *The tacit dimension.* New York: Anchor Books.

Randall, R., & Brown, A. (2015). *In time for tomorrow? The carbon conversations handbook.* Stirling: Surefoot Effect.

Venn, C. (2018). *After capital.* London: Sage.

Westcott, G. M. (2016). *The role of subjective factors in local authorities' action on climate change in South West England.* Ph.D., University of the West of England. Available from: http://eprints.uwe.ac.uk/28966.

Winnicott, D. W. (1991). *Playing and reality.* London: Routledge.

Part II

Mostly Findings

This part of the book focuses on findings from five studies which have all used qualitative and psycho-social influenced methodologies but which emphasise the findings generated by the research rather than methodologies deployed.

Jo Hamilton provides a much needed survey and initial evaluation of a range of civil society initiatives which offer containment to those first developing an awareness of the implications of climate change and to those, like Sally Gillespie's participants, who are more deeply engaged with these issues. The aim of all these initiatives is to provide participants with a vehicle for working through the emotional and ethical challenges they face and thereby free up their capacity to think and act in ways which are appropriate for them. The initiatives do this by drawing on a variety of resources, including psychotherapy, mindfulness, nature connection, yoga, etc. and also the power of the group. In her interviews with the facilitators of these initiatives Hamilton uses many of the methods for eliciting narrative and of deep listening that we encountered in the Part I of this book.

In his contribution Rembrandt Zegers returns to the theme of different ways of knowing which was touched on by several earlier contributors to this collection. Zegers makes the case for sensory experience as a way of knowing, one that reconnects us to the bodily foundations of experience and therefore to the ways of knowing of the other-than-human. Zegers shows how the capacity to trust our senses is not only important in the research process itself but, through three fascinating case studies, he illustrates how this capacity has informed the work of individuals who have learnt from nature (from horses, cows and glaciers) and used this learning to lead others.

In the introductory chapter mention was made of recent survey data which indicated that whilst most people know about climate change, and even acknowledge the human element to it, they remain relatively indifferent to it. The contributions by Gill Westcott and Robert Tollemache focus very much upon the experience of this knowledgeable but relatively indifferent chunk of the population. Westcott undertook detailed interviews with local government politicians and officers in the UK at a time when post-crash austerity measures had effectively ended many of the sustainability initiatives that had developed under the previous Labour government. Following Kari Noorgard's work, Westcott looks at the way in which denial becomes embedded in the warp and weft of everyday organisational life. She notes the conversational gambits—flippancy, cynicism, defensiveness, diffusion of responsibility—her respondents would deploy to reconcile their awareness of climate change with conformity to the social norms and expectations of 'business as usual'. Westcott's research also throws interesting light on the relation between awareness and action. Our common sense is that by increasing people's awareness of the issue you then enhance their propensity for action. Westcott questions this. She finds evidence for the reverse; that an increased sense of one's own self-efficacy (one's capacity for taking action) leads to a greater capacity to contain difficult information and feelings.

Tollemache's research offers a detailed snapshot of London's middle class. The lawyers, teachers, theatre workers and academics in his sample manifest complex and contradictory thoughts and feelings about climate change. His interviews help to unpick the concept of denial and provide many useful illustrations of negation, disavowal, and literal, implicatory and interpretive denial. Interestingly, many of his respondents also talked about Stan Cohen's fourth form of denial, one he referred to as 'silence', that is, the pressure not to talk about the elephant in the room, and the real or imagined sanctions that are applied to 'silence breakers'. As with Westcott, Tollemache's respondents seemed very aware of social pressures and the power of like-mindedness but Tollemache also notes the importance of one's individual life history and the impact of early childhood environments on the adult's experience of nature.

The final chapter in this part takes us back to the experiences of what Sally Gillespie had earlier referred to as the 'canaries' of climate change, in this case climate scientists and activists who are immersed in the issue every day. Rosemary Randall and Paul Hoggett note how the contrasting cultures of science and activism leave their participants differentially equipped to deal with the emotional and ethical challenges thrown up. Jo Hamilton's chapter has already revealed some of the many initiatives designed to enhance emotional literacy and self-awareness within activist millieux and the experiences of the activists interviewed by Randall and Hoggett reveal some of this emotional reflexivity. In contrast the scientists talked of an academic culture which seemed ill equipped to manage these challenges and used 'social defencesSocial defences' such as the idealisation of scientific rationality and neutrality to manage them. This then left them ill prepared for engagement with publics and policy makers, an engagement which was then left to a minority who often felt isolated, distrusted and under attack. Randall and Hoggett speculate that this may then foster caution in those from the scientific community who do engage leading them to 'pull their punches' and avoid anything that might be construed as 'alarmism'.

8

Emotions, Reflexivity and the Long Haul: What We Do About How We Feel About Climate Change

Jo Hamilton

Introduction

A step change in responses to climate change is needed. There is scientific consensus about the causes and impacts of climate change, but in high per-capita carbon emitting countries such as the UK, USA and Australia, mitigation action is not prioritised, insufficient, and lacking in ambition and resource. A radical rethink is needed to bring about just and equitable mitigation and carbon emission reductions, and to adapt to the physical, political and psychological implications of a warming world.

The range of emotions[1] triggered by climate change has implications for how individuals, organisations and societies respond. In western societies, these emotions include fear, helplessness, guilt,

---

[1] I am using emotions as the main descriptor for emotions, affects and feelings, and will distinguish between them where necessary.

J. Hamilton (✉)
Department of Geography and Environmental Science, School of Geology, Archeology and Environmental Science, University of Reading, Reading, UK

© The Author(s) 2019 **153**
P. Hoggett (ed.), *Climate Psychology*, Studies in the Psychosocial,
https://doi.org/10.1007/978-3-030-11741-2_8

anxiety, potential loss, grief, trauma, processes of anticipatory mourn-
ing and anger, solastalgia (e.g. Norgaard 2011; Stoll-Kleemann et al.
2001; Lertzman 2015; Randall 2009; Head 2016; Doppelt 2016;
Weintrobe 2013; Albrecht et al. 2007) and 'environmental melancholia' (Lertzman 2015).

Emotional responses to climate change are influenced by many factors,
such as perceptions and experiences of agency. Research with climate scientists and activists in the UK (i.e. those with connection to forms of
agency) suggests that there is an *undercurrent of trauma, despair and
defensive coping* evident in both sectors (Hoggett and Randall 2016).
It is sometimes difficult to acknowledge or express the emotions associated with climate change. Yet without permission or opportunity for
expression, individuals and societies can remain in a state of emotional
paralysis, which can inhibit engagement, action, and a response-ability
(Norgaard 2011; Lertzman 2015; Macy and Brown 2015).

Broader engagement approaches are necessary, which can acknowledge
emotions associated with climate change and enable latent concerns to
surface (Hobson 2008; Moser 2016; Lertzman 2015; Head 2016). Such
approaches could provide space, permission and safety to work through
difficult and complex emotions, and *'find creative means of allowing a
full range to be present'* (Lertzman 2015, p. 13). Taking this psychosocial
approach involves translating some practices, which are primarily done
in private psychotherapeutic practice at present, into broader social contexts. This requires a *'collaboration between psychotherapists, social scientists, and communicators'* (Lertzman and Norgaard 2011, p. 9).

Adams (2016) discusses the need for research (and, I would add,
interventions and action) which can expand, not foreclose, the possible range of climate change narratives and responses. Such interventions could involve *'fostering imaginative capacities in supportive contexts'*
(2016, p. 245), including different modes of communication, and
opportunities to experience affectual and embodied responses to climate
change, and re-connecting with the more than human world.

These expansive approaches to climate change engagement could
involve 'emotional reflexivity'. By 'emotional reflexivity' in this context,
I mean developing an embodied and relational awareness of how people engage with and feel about an issue, how this influences the actions

they take, the stories they inhabit and perceptions of their individual and collective change and agency (Brown and Pickerill 2009; Holmes 2010). Such approaches can enable a range of emotions to be present, and disrupt 'narrative foreclosure'. This approach to emotional reflexivity acknowledges that emotions are mutable, and move, change, transform and can be transformative when they have the space and safety to do so (Ahmed 2014).

Emotional reflexivity in the context of social and environmental action has been discussed by Brown and Pickerill (2009), King (2005) and Gould (2015), amongst others. King's research into the use of Re-evaluation Co-Counselling by social change activists in Australia found that the practice enabled participants to cope with values which were in opposition to mainstream society, to '*free up her capacity to think*' and enable more creative responses (2005, p. 161). Both Brown and Pickerill (2009) and King (2005), highlight the need for activists and social change organisations to incorporate emotional reflexivity into their activities and urge for more research about emotional reflexivity in social change contexts.

This chapter presents a variety of 'Emotionally Reflexive Methodologies' used in grassroots and civil society organisations in the UK. It provides an overview of existing research into ERMs, followed by presentation and discussion of new research about the relevance of these ERMs to engaging with climate change, and some of the challenges posed.

Emotionally Reflexive Methodologies

There is a range of methodologies which facilitate an acknowledgement, processing, and potential transformation of emotions around social and environmental issues. These can collectively be termed 'Emotionally Reflexive Methodologies' (hereafter ERMs). Whilst differing in context, lineages drawn on, accessibility and the timescales they are conducted over, they all involve a degree of emotional reflexivity. Positioned at the interface of psychological and social approaches to climate change, ERMs are prime areas for psychosocial research. Selected ERMs which incorporate reflexivity about climate change are presented in Table 8.1, together with a brief overview of relevant research.

Table 8.1 A selection of Emotionally Reflexive Methodologies relevant to climate change engagement

Name, website, reference	Brief description	Research summary
The Work That Reconnects (TWTR) Macy and Brown (2015) https://workthatreconnects.org/ '*Active Hope*: How to face the mess we're in without going crazy' (Macy and Johnstone 2012) https://www.active-hope.info/	Groupwork processes developed in 1980s by Joanna Macy and colleagues Draws on system theory, Buddhist philosophy and deep ecology, with focus on holistic connection to life *Active Hope*: draws on TWTR cycle, groups are facilitated to work through the book together	• Johnstone (2002) conducted a follow up survey to workshop participants, and Prentice (2003) and Hollis-Walker (2012) reflected on workshops • Majority of participants found experience 'personally healing' (Johnstone 2002), a minority of participants experienced negative impacts alongside this (Johnstone 2002) • Participants deepened connections to self, others and the more than human world • Hollis-Walker (2012) and Johnstone (2002) noted the renewed commitment to action
Inner Transition (I.T.) https://transition-network.org/do-transition/inner-transition/	I.T. is a core component of the Transition Network, the international network of the Transition movement. It covers a variety of practices offered as stand-alone workshops, or incorporated as part of group culture	• I.T. practices contributed to development of successful projects; encouraged emotional awareness in all activities (Banks 2012), and developed literacy around 'parallel processes' within Transition groups (Prentice 2012, p. 186) • The degree of integration or polarisation between I.T. and more practical aspects of Transition Initiatives were noted by Ruchetto and Poland in Canada (2015), Power in Australia (2016), and Banks (2012)

(continued)

Table 8.1 (continued)

Name, website, reference	Brief description	Research summary
Carbon Conversations http://www.carboncon-versations.co.uk/	Conducted through groups of 6–12 people, who meet for facilitated meetings, and work from a book (Randall and Brown 2015) The approach *'addresses the practicalities of carbon reduction while taking account of the complex emotions and social pressures that make this difficult'* (http://www.carbonconversations.co.uk/)	• Carbon Conversation groups provided structure to support the emotional responses to climate change (Randall 2009) Key conclusions following surveys and interviews with participants (Büchs et al. 2015) were: • Sharing experiences helped participants become *aware of* and reflect on their feelings and inner conflicts • Group dynamics affected participants' capacity to do this • Supported participants to take carbon reduction action • Works best for those on cusp of change
Carbon Literacy Project (CLP) http://www.carbonlit-eracy.com/	Initiated in Manchester, the Carbon Literacy Project (CLP) *'offers everyone who works, lives and studies in Greater Manchester a day's worth of Carbon Literacy'*, and supports participants to take action in their workplaces and communities	• Internal and external research has been conducted on the CLP (http://www.carbonliteracy.com/research/) • Reports attested to increased motivation and agency to take action (Richards 2017) and the range of political engagement (Moore 2017)

(continued)

Table 8.1 (continued)

Name, website, reference	Brief description	Research summary
Mindfulness based interventions	Mindfulness and Behaviour Change programmes delivered in UK to behaviour change practitioners (Lilley et al. 2016, Whitehead et al. 2017)	• Large evidence base on the effectiveness of mindfulness throughout society (e.g. Mindful Nation UK report 2015; Barrett et al. 2016) • Evaluation from 'Mindfulness, Behaviour Change and Engagement in Environmental Policy' participants (Lilley et al. 2016) noted increased understanding of role of emotions, values and norms in decision making. Evaluation from Mindfulness and Behaviour Change programmes concludes that these approaches 'open up interesting opportunities for conceiving more empowering and ethically sensitive approaches to behavioural government' (Whitehead et al. 2017, p. 133)
ERMs within settings of social and environmental action ERM workshops in activist spaces	Range of approaches offered, including TWTR workshops, trauma support, meditation and wellbeing support offered in activist spaces such as Reclaim the Power, and anti-fracking camps	Hoggett and Randall (2016) mention social support systems within direct action movement Current research by the author reflects on some facilitator experiences of offering workshops in places of protest

(continued)

Table 8.1   (continued)

Name, website, reference	Brief description	Research summary
Social Permaculture MacNamara, L. (2012) **Earth Activist Trainings (EAT)** https://earthactivist-training.org/	The Permaculture principles of have been adapted and expanded to incorporate a focus on establishing social relations and human culture, such as Social Permaculture (McNamara 2012) EAT courses combine 'permaculture, earth-based spirituality, organising and activism'	Puig de la Bellacasa (2010) reflected on positive practical and affective outcomes: *'the affect cultivated in Earth Activist Trainings is not despondency in front of the impossible, but joy in the hope of possibility'* (2010, p. 162)

Table 8.1 gives examples of ERMs practised in the UK. My selection criteria included ERMs framed by environmental and social action; mainly conducted in person; and which encourage reflexivity about climate change. Those selected are mainly verbally based: i.e. primarily but not exclusively through talking, reflecting and sharing in face-to-face group settings. These ERMs comprise a collection of exercises, ceremonies, rituals and approaches, which can be experienced together (for example a cycle of 'The Work that Reconnects'), or as practices in wider contexts (such as mindfulness exercises and opportunities to explore feelings in meetings and conferences).

There are many other practices incorporating emotional reflexivity which are used by individuals and groups. Alongside faith-based approaches, and online courses, these can and do support engagement and action with social and environmental issues but are not necessarily *framed* by them. These include mindfulness, meditation, yoga, nature connection, creative practices, embodiment and movement practices. Additionally, many ERMs are incorporated into trainings and facilitation offered by collectives and centres which support social and environmental action (e.g. Navigate, Seeds for Change and Tripod Training,[2]

---

[2]See references for details of these groups.

Embercombe in Devon, EcoDharma in the Catalan Pyrenees). These wider practices and places of emotional reflexivity are beyond the scope of my current research but would be valuable to explore in future research and could offer rich insights for application in a range of progressive social contexts.

Development and Lineages Drawn on by ERMs

These ERMs have evolved, changed, and been used in different contexts over at least four decades. However, as shown in Table 8.1, with the exception of Mindfulness there is a dearth of existing and accessible research about the relationship between participating in these ERMs and engaging with climate change and related issues. That which exists primarily focuses on single ERM cases, and attests to the positive impacts of participating, and the potential for wider application (e.g. Randall 2009; Johnstone 2002; Büchs et al. 2015).

The ERMs draw on a range of 'lineages', including Western psychology, Eastern philosophies and spiritual practices, and indigenous social practices and wisdom traditions (Prentice 2012). They include grief work (Randall 2009; Weller 2015) psychoanalytic theories and group work (Stock Whitaker 2001; Bion 1961), ecopsychology (Rust 2008), trauma (Doppelt 2016; Herman 1992), faith based approaches such as Buddhism (Macy and Brown 2015), nature connection, deep ecology (Seed et al. 1988) feminist theory and practice, systems theory, and indigenous practices (e.g. grief tending).[3] The range of approaches used in any one ERM workshop will depend on the facilitator's experiences, practices, skills and trainings, and crucially, the goals and context of the workshop: to whom and where they are offered.

---

[3]Practices that facilitators have mentioned grief tending which draws on Dagara traditions from Burkina Faso as taught by Malidoma Somé (http://malidoma.com/main/) and the late Sobonfu Somé (http://www.sobonfu.com/).

Research Methods and Data

My research is informed by trans-disciplinary theoretical perspectives, including human and cultural geographies (Bondi 2005; Head 2016), sociology and cultural politics (Norgaard 2011; Ahmed 2014), and psychosocial perspectives (Weintrobe 2013; Adams 2016; Lertzman 2015; Hollway and Jefferson 2000).

Primary data collection was conducted between February 2017–February 2018. This included online searches of ERMs using key words,[4] and semi-structured interviews and ongoing conversations with 28 facilitators of different ERMs. The interviews were transcribed verbatim, followed by a second listening to observe non-verbal aspects, and my felt sense, of the interviews. The interviews were informed by psychosocial methodologies (e.g. Hollway and Jefferson 2000), such as asking for the *story* of how they came to be facilitating their methodology, and following the narrative flow of the interviewee, but weren't using psychosocial interview methodologies per se. For the data presented here I have primarily relied on the 'told story' of the interviewees.

This chapter focuses on the facilitator interviews. It is also informed by participatory aspects of my research which included participating in a range of ERMs, re-training to facilitate *The Work that Reconnects*, and keeping a reflective diary. This will be complemented by my current ongoing research which explores participant's perspectives of ERMs.

My positionality will have informed my access to facilitators and my analysis. I am a white, middle class woman from the UK. In the past

---

[4]Key words included names of methodologies such as 'Inner Transition', 'The Work that Reconnects' and related practices. Searches were also conducted in online groups, such as 'The Work that Reconnects UK' facebook group (https://www.facebook.com/groups/111468605594102/), and Ecopsychology online group (http://ecopsychologyuk.ning.com/).

two decades, my work has involved environmental campaigning, group facilitation, action and research about community energy projects and low carbon community groups. I have found personal value in participating in a variety of ERMs and during this research have stepped back into facilitating *The Work that Reconnects* and campaigning on climate change issues.

What ERMs Can Achieve

This section gives a brief overview of what ERM *facilitators* experienced and observed occurring in the ERMs they facilitate. It presents and discusses the emotionality of climate change observed in the ERMs, and how ERMs can contribute to engaging, and maintaining an engagement, with climate change. It then reflects and discusses how ERMs can create the conditions for emotions to be safely acknowledged, expressed and worked through.

Emotionality of Climate Change

In the ERMs, facilitators observed a range of emotions connected to the issues and implications of climate change. The observed emotions were mainly congruent with the literature cited above, such as grief and loss, sadness, shame, anger, fear, guilt, frustration, anger, overwhelm and despair. For example:

> "*What I meet first is grief, and then often, less, inner paths going 'oh forget it, it's too big, who am I? oh anyway it's not possible, I can't do it*" [F13:7]; and

> "*how people deal with their feelings… you know despair is always there in a sense… we're all in a process together… which is a kind of falling apart. As well as a new story emerging, but that's so small isn't it, compared to the falling apart that is happening*". [F10:7]

However, more positive, connecting and action-oriented emotions were also present, such as empathy, hope and gratitude. These positive emotions mainly occurred as an outcome of the reflexivity and *working through* (not avoiding) painful emotions. Alongside this, interconnectivity was encouraged through opportunities to experience it within (one's own emotions), between (with other participants), and beyond (with the more than human world). As well as developing safety for participants, interconnectivity can in turn enable an opening to wider resources, such as developing an inner resilience and agency to act with others.

In some cases, the positive emotions were reflected through the 'framing'[5] of the ERM. For example, the *Carbon Literacy Project* aims to facilitate an active engagement with climate change, and for many participants may be their first workshop on the topic. The frame is hopeful and action oriented, whilst the interactive nature of the workshops enables acknowledgement and recognition of individual and collective agency (see Table 8.1). One of the initiators explained: "*do it in a frame that is about hope. This is about how we, the many different 'wes' are going to sort it. The hope in action, the positivity in action*" [F11].

It is clear that, at least initially, emotions linked to climate change are linked to other related issues such as anger at insufficient political responses. For example:

> "*Anger, 'how did we ever allow this to get to this stage?' 'Why aren't people acting when it's so bloody obvious that we've got to?'. These are the things that will come up*". [F5:15]

Facilitators observed that through the acknowledging and working through of difficult emotions, the threads can be teased out and re-integrated to enable agency. For example, one facilitator reflects that *The Work that Reconnects* is: "*good at kind of bringing the intellectual understandings together with the sort of emotionality, and bringing awareness to the emotionality and being able to harness the energy of it*" [F15].

---

[5]Framing in its simplest form means how an issue is contextualised, and the concepts and issues it links to. There is much research and discussion on framing climate change and action, see Lertzman (2015).

Engaging and Taking First Steps

ERMs have been used to help people actively engage in the issues and implications of climate change. Facilitators observed that participants can work with and generate a reflexivity around their responses—emotional or otherwise—to climate change, as long as there is space and permission to name the arising emotions, and that they are held. This is reflected by the initiator of *Carbon Conversations*:

> *"people would often come out of the groups feeling much better equipped, knowing what they wanted to do, feeling they were part of something … positive things were that people came out feeling, feeling … better about the things which frightened them, feeling better about what they could do, feeling they were part of something".* [F6]

Participation can also be a trigger to becoming involved in local action, as a *Work That Reconnects* facilitator mentioned:

> *"as a result of that work, she just got completely switched on to environmental action and joined Transition in her area, and just hasn't stopped".* [F15]

How the ERMs relate to *practical* action on climate change is an important consideration. For example, integrating TWTR into a permaculture course was well received. The group had formed, relationships had been established, and it was framed within *"a solution focused approach, [which encouraged] positive mental attitudes"* [F14].

A facilitator who was key in establishing TWTR within a Buddhist community, reflected that doing TWTR was a *"condition that brought… into being… [our] willingness to go from a point of … eyebrows [being] raised, at one order gathering …[in] 2007, that we were even doing a day on climate change… through to the position now where we're very active in Buddhist action month in June every year"* [F15].

At the engagement end of the spectrum, it seems that ERMs contain clear examples of processing the issues and implications of climate change. However, the relationship between ERMs and action on climate change will also be influenced by involvement, or not, in a community of practice (Wenger 1998), and the culture of that community. Some facilitators mentioned that they had encountered defences to some ERMs or practices, which were pejoratively referred to as 'touchy feely stuff', or which were seen as superfluous to taking action.

Resourcing Activists and Sustaining Action

Burnout and exhaustion is very common for those involved in taking forms of environmental action, whether at community level or campaigning and non-violent direct action (Hoggett and Randall 2016). This is caused by, amongst other things, dealing with the knowledge of the climate impacts occurring already, the scale of transformation needed, and inadequate political responses. Additionally, there can be discouragement from the wider public, and for some involved in direct action, being traumatised through the actions of the state or security forces.

Facilitators observed that participants appreciated a place where they could show up and inhabit a different mode to the norms of meetings of events, where emotional expression is largely absent. The following quotes give examples of these different modes of being together, alongside an acknowledgement of ways of working with distress which can build connections within and between participants:

*a space to be real together, a space to be honest, to be vulnerable if that's where they are... to kind of stop, to **be** a bit* [F9, my emphasis];

to actually be with, to turn towards ... what is really going on for you ...it's not going to do you in, or there are ways of working with your actual experience, you don't have to run away from it, or say it's not happening, or explain it away, or feel bad about it [F15];

In the 'activist environmentalist' sort of area, it tends to be people who are just slogging along and feeling really really driven, and coming into that situation of sitting in a circle with others is just a sharing of their load, just to feel they're not alone, just really lightens the load...it's the release and relief bit, and also a connection bit to get from isolation to 'god I don't have to save the world on my own'. [F14]

Having a space to consider the longer timescales over which action takes place was also recognised as important, as one facilitator reflected:

> *for me it's about the timescales that we're willing to deal with change in....it takes time to work through that process, and it takes time for the infrastructure to change to support that, ... part of the Inner Transition ... it's that... feeling your pain for the world, and mourning you know what's lost, and that inter-generational timescale of connecting with the ... seven generations time... it's a sort of reframing exercise almost.* [F25]

The quotes above build on evidence presented in Table 8.1. They provide examples of one-off workshops, and ongoing practices and groups, whereby *permission* is given to acknowledge emotions, within a supportive context that links inner work with outer action. These examples are often complemented by nature connection practices informed by ecopsychology, deep ecology and indigenous wisdom traditions, which can acknowledge and help build resourcing relationships to the more than human world.

Holding and Containing Emotions in Space and Time

Different ERMs can open up and work with differing degrees of depth and strong emotions around social and environmental issues. This can present challenges for facilitators, as participants will have differing experiences of, and comfort with, workshop settings, exploring and embodying their own emotions, and witnessing other people's emotions. An awareness and care about how to contain and hold the range of emotions in connection with others, was evident throughout the facilitator interviews. This section discusses notions of *containment* (drawing on Stock Whitaker 2001), which is of importance when considering how some psychotherapeutic practices are used in a range of group and semi-public settings. It is presented here to stimulate further discussion and reflection on how best to support and develop these ERMs.

Depending on the ERM used, facilitators may start with exercises which can open-up emotional responses, together with exercises that help ground and contain, and connect to resources within, between and beyond the individual. For example, the first stage of *The Work that Reconnects* includes exercises to encourage an exploration and experience of gratitude and interconnectedness.

Building a group to contain the more painful or challenging emotions is done in stages. Solo or pair work is an important component of this, for example:

> *what I tend to rely on …is people's …quiet exploration of their own, of what's going on inside them, maybe in a small group like a pair … where there's plenty of time around it, where expression isn't actually required, but the kind of um curiosity and exploration of their own inner processes is required… Then they come back into a circle and by that time they feel safe. And then it's OK for them to actually express something in the group.* [F3]

Exercises which incorporate creativity can also enable some emotions to be accessed, as they are held in forms such as drawing, writing or storytelling. For example:

> *creative things like … inviting people to just take a pen and a piece of paper, and then in response to a question have some time alone drawing, and then explain their drawing to their paired partner or something.* [F17]

Holding can also occur through fostering a deeper connection with nature and the more than human world. For example, this facilitator combined poetry, indigenous wisdom traditions and nature connection:

> *we did these poems, and we listened, and we encouraged people to go out, and just wander by this burn[6] … to just see what the burn has to say to you. Observe it with all your senses … if you use your owl vision, where you soften your eyes, and you can see all around you almost, it wakes up all your senses more … And it was amazing what people came back with, you know having*

---

[6]'Burn' in this case is a stream in Scotland.

gone out, then came back with, and they shared in pairs, and then they shared as a whole group ... And people on that day retreat went deeper than sometimes people go in a week. [F7]

Exercises where strong emotions are often expressed are approached with care and reverence. One experienced facilitator of *TWTR* mentioned that "*before holding such things like [The Truth Mandala⁷] you have to teach people the grounding methodologies to enable them to safely enter it*" [F5]. These include breathing, mindful awareness of emotions, and practices which can connect to wider resources, for example a physical object, or an image of a safe place. She continues:

I've always had that concern that [The Truth Mandala exercise] needs to be held very safely ... it feels like one thing that's needed to make it safe is to have a very deep understanding of the process, of what it is, and from that have a very clear understanding of why the instructions you give are very important, every one of them. [F5]

A number of facilitators mentioned the importance of creating ceremonies or rituals to help contain stronger emotions. Some ceremonies or rituals draw on interfaith or shamanic trainings. For example:

For me personally I'm increasingly pulled towards a more spiritual approach, and learning how bringing in a level of ceremony into almost everything that I do brings a wider awareness and connection and involvement with things that we can't see and things that we don't quite know about, that actually make things work better [F14]; and

[I draw on trainings which] teach you how ....to create physical and energetic space that is holding, and bounded and contained with a beginning and with an end, and a closing. [F5]

Containing is also invoked across time and space, for example through dedicating the intention of some exercises to '*the welfare of all beings and*

---

⁷The 'Truth Mandala' is a specific exercise in The Work that Reconnects, which enables participants to explore and express pain for the world (Macy and Brown 2015, p. 121).

the healing of our world' (Macy and Brown 2015, p. 121), and through inviting other people, and the more than human world, to hold the facilitators and workshop participants.

What participants do after workshops or courses is pertinent to continuing and integrating the experiences and learnings from the ERM. Some ERMs have been criticised for opening a space for emotional expression, without necessarily equipping participants to deal with them reflexively, and on an ongoing basis. This is particularly relevant for stand-alone ERMs which take place over a day or weekend and may involve people who do not know each other prior to a workshop, or see each other after it. Shorter workshop lengths (e.g. a few hours) can be insufficient timescale to enable more painful emotions to surface, be contained and be adequately processed. Recognising this, most facilitators will offer tools and resources at workshops, to enable participants to continue, sustain, and develop practices and inquiries introduced in the workshop.

Ongoing ERMs, for example meeting over a period of 6 weeks, or monthly ongoing meetings, can enable a gradual building of a group and development of practices. These practices can include mindfulness, nature connection practices, and journal writing. For example, *Active Hope* (Macy and Johnstone 2012) is usually experienced over 4–6 weeks, enabling cycles of action, reflection and group discussion. Facilitation can be shared, which can encourage plural approaches and styles, and collapse the 'facilitator / participant' dynamic. Some facilitators have integrated approaches. For example, *Active Hope* has been combined with Mindfulness, which supports participants to increasingly explore and build up their capacity to be with the more painful emotions.

In these ERMs, tensions were observed between offering a space for emotional reflexivity which can build up the capacity to act without being linked to particular ways of action, (developing agency) and ERMs which are directed towards supporting certain forms of action or advocacy (addressing the urgency). This has been recognised (e.g. Lertzman 2015; Lilley et al. 2016) and remains an ongoing challenge. Creatively navigating and negotiating those tensions to ensure that connections are made between the inner and outer dimensions of action could involve building individual and group agency and capacity to take action on climate change and ensuring there are a range of opportunities to enact the agency.

Communities of practice, such as spiritual or workplace communities, can enable the integration of experiences, and increase the likelihood of collective action. For example, an initiator of the Quaker 'Living Witness' programme mentioned that:

> the events themselves are kind of trigger points. I don't see the transformation happening that much in the weekends. I mean I think it's more the people who … [have] been working with the same community over fifteen years. [F2]

In this way, conversations about climate change can be broadened to include an emotional dimension, alongside encouraging practical action to reduce CO_2 emissions at personal and organisational levels.

Discussion and Conclusions

This chapter has mapped out some ERMs used in the UK and beyond. The ERMS offer a variety of ways to be with the implications and emotions of climate change; to explore them at cognitive, somatic and emotional levels; and to help facilitate a grounded, active and resilient engagement. These ERMs are places where the inner psychological and embodied dimensions meet and dance with the outer engagement and action on environmental and social action. They combine practices as old as the hills, with the issues which threaten the very ground we stand on.

These ERMs, which have been primarily explored in grassroots and civil society organisation settings and communities of practice, offer a range of methodologies and exercises which could be used and adapted to suit different contexts, and at different scales. For example, the *Carbon Literacy Project* integrates emotional reflexivity into workplace situations to support engagement with climate change; *The Work That Reconnects* can be used to explore deeper emotions to enable action to be sustained; and *Inner Transition* involves a range of practices support the culture, effectiveness and reach of Transition initiatives. There are many opportunities for learning between ERMs, to explore how they can be effectively used in different sectors, at different scales and to varying degrees of emotional depth. To enable this wider application however, there needs to be greater visibility of the ERMs, and a clearer link between individual and social resilience, and action on climate change.

ERMs are located between disciplines. They are not private therapeutic spaces, although clearly therapeutic activities can occur within them. They are held by facilitators who are trained in different methods, but not necessarily therapeutically trained. Facilitators are encouraged to reflect on their experiences and competencies, for example *The Work That Reconnects* has a 'Facilitator Competency Framework',[8] but there is clear potential for learning, development, peer-to-peer mentoring and cross-fertilisation of the creative edge between different skillsets of ERM and therapeutic lineages.

Through discussing a range of ERMs, this chapter has expanded existing research in this area and explored the potential that ERMs offer for engaging and sustaining an active engagement with climate change. However, it also raises many questions and opportunities for further research. These include how to safely open and contain spaces for emotional reflexivity in different contexts and in different cultures? How do experiences of trauma and grief intersect with privilege, power and everyday experiences of climatic change? What are the risks of doing so, and of not doing so (Randall 2009)? How could the sharing of support, insights and experience between disciplines and ERMs help to develop them?

At the time of writing, the UK and northern hemisphere have experienced record breaking temperatures causing wildfires, deaths, and water shortages. This is alongside a volatile and increasingly political environment with depleted physical and social infrastructure to deal with and respond to the intersecting challenges posed by climate change. This brief overview of ERMs contains a range of practices which could help

---

[8]Work that Reconnects Facilitator Competency Framework [online]. Available from: https://workthatreconnects.org/wtr-facilitator-competency-framework/ [accessed 20 June 2018].

the public, and those involved with, or on the cusp of, taking action on climate change, to get involved, stay involved and disrupt the stagnant range of responses on offer from governments. Climatic tipping points are already occurring, the time is ripe for tipping points in how we respond to these issues. The ERMs presented can offer ways to enable and sustain an engaged response in these uncertain times.

Acknowledgements I would like to thank all the facilitators who were interviewed, for their contribution of time and expertise, and to my supervisors Hilary Geoghegan and Giuseppe Feola. I would also like to thank those who commented on earlier drafts: Paul Hoggett, Rosie Robison and Joe Butler.

References

Adams, M. (2016). *Ecological crisis, sustainability and the psychosocial subject: Beyond behaviour change.* London: Palgrave Macmillan.

Ahmed, S. (2014). *The cultural politics of emotion* (2nd ed.). Edinburgh: Edinburgh University Press.

Albrecht, G., Sartore, G.-M., Connor, L., Higginbotham, N., Freeman, S., Kelly, B., et al. (2007). Solastalgia: The distress caused by environmental change. *Australian Psychiatry, 15* (Supplement). S95–S98.

Banks, S. (2012). Inner Transition survey results. *Transition Network.* http://www.transitionnetwork.org/resources/inner-transition-survey-results. Accessed 12 June 2016.

Barrett, B., Grabow, M., Middlecamp, C., Mooney, M., Checovich, M., Converse, A., et al. (2016). Mindful climate action: Health and environmental co-benefits from mindfulness-based behavioral training. *Sustainability, 8,* 1040. https://doi.org/10.3390/su8101040.

Bondi, L. (2005). Making connections and thinking through emotions: Between geography and psychotherapy. *Transactions of the Institute of British Geography, 30*(4), 433–448.

Bion, W. R. (1961). *Experiences in groups and other papers.* London: Tavistock.

Brown, G., & Pickerill, J. (2009). Space for emotion in the spaces of activism. *Emotion, Space and Society, 2,* 24–35.

Büchs, M., Hinton, E., & Smith, G. (2015). 'It helped me sort of face the end of the world': The role of emotions for third sector climate change engagement initiatives. *Environmental Values, 24*(5), 621–640.

Doppelt, B. (2016). *Transformational resilience: How building human resilience to climate disruption can safeguard society and increase wellbeing*. London: Routledge.

Gould, D. (2015). When your data make you cry. In H. Flam & J. Kleres (Eds.), *Methods of exploring emotions* (pp. 163–171). Abingdon, Oxon: Routledge.

Head, L. (2016). *Hope and grief in the anthropocene: Reconceptualising human nature relations*. London: Routledge.

Herman, J. L. (1992). *Trauma and recovery: The aftermath of violence—From domestic to political terror*. New York: Basic Books.

Hobson, K. (2008). Reasons to be cheerful: Thinking sustainably in a (climate) changing world. *Geography Compass, 2*(1), 199–214.

Hoggett, P., & Randall, R. (2016). *Sustainable activism: Managing hope and despair in social movements*. Open Democracy.net. Available from: OpenDemocracy.net/transformation/paul-hoggett-rosemary-randall/sustainable-activism-managing-hope-and-despair-in-socia. Accessed 12 June 2018.

Hollis-Walker, L. (2012). Change processes in emotion-focused therapy and the work that reconnects. *Ecopsychology, 4*(1), 25–36.

Hollway, W., & Jefferson, T. (2000). *Doing qualitative research differently: Free association, narrative and the interview method*. London: Sage.

Holmes, M. (2010). The emotionalization of reflexivity. *Sociology, 44*(1), 139–154.

Johnstone, C. (2002). Reconnecting with our world. In A. Chesner & H. Hahn (Eds.), *Creative advances in groupwork* (pp. 186–216). London: Jessica Kingsley Publishers.

King, D. (2005). Sustaining activism through emotional reflexivity. In H. Flam & D. King (Eds.), *Emotions and social movements* (pp. 150–169). London: Routledge.

Lertzman, R. (2015). *Environmental melancholia*. London: Routledge.

Lertzman, R., & Norgaard, K. (2011). A dialog between Renee Lertzman and Kari Norgaard. *Ecopsychology, 3*(1), 5–9.

Lilley, R., Whitehead, M., Howell, R., Jones, R., & Pykett, J. (2016). *Mindfulness, behaviour change and engagement in environmental policy*. Aberystwyth University. Available from: https://changingbehaviours.files. wordpress.com/2016/01/mindfulnessreportgapfinal.pdf. Accessed 12 Apr 2018 [Online].

MacNamara, L. (2012). *People and permaculture*. East Meon, Hampshire, UK: Permanent Publications.

Macy, J., & Brown, M. (2015). *Coming back to life. The updated guide to the work that reconnects*. Gabriola Island, BC, Canada: New Society Publishers.

Macy, J., & Johnstone, C. (2012). *Active hope: How to face the mess we're in without going crazy*. Novato, CA: New World Library.

Mindfulness All-Party Parliamentary Group (MAPPG). (2015). *Mindful Nation UK report*. Available from: http://www.themindfulnessinitiative. org.uk/images/reports/Mindfulness-APPG-Report_Mindful-Nation-UK_ Oct2015.pdf. Accessed 12 June 2018.

Moore, K. (2017). *(Re)Politicising climate change engagement: A case study of the Carbon Literacy Project*. Masters dissertation [online]. Available from: http:// www.carbonliteracy.com/research/. Accessed 12 June 2018.

Moser, S. (2016). Reflections on climate change communication research and practice in the second decade of the 21st century: What more is there to say? *WIRES Climate Change, 7*, 345–369.

Navigate Co-operative. https://navigate.org.uk/. Accessed 2 June 2018 [Online].

Norgaard, K. M. (2011). *Living in denial: Climate change, emotions and everyday life*. Cambridge, MA: MIT Press.

Power, C. (2016). The integrity of process: Is inner transition sufficient? *Journal of Social and Political Psychology, 4*(1), 347–363. https://doi. org/10.5964/jspp.v4i1.538.

Prentice, H. (2003). Cosmic walk: Awakening the ecological self. *Psychotherapy and Politics International, 1*(1), 32–46.

Prentice, H. (2012). 'Heart and Soul': Inner and outer within the transition movement. In M.-J. Rust & N. Totton (Eds.), *Vital signs* (pp. 175–190). London: Karnac Books.

Puig de la Bellacasa, M. (2010). Ethical doings in naturecultures. *Ethics, Place and Environment, 13*(2), 151–169.

Randall, R. (2009). Loss and climate change: The cost of parallel narratives. *Ecopsychology*. https://doi.org/10.1089/eco.2009.0034.

Randall, R., & Brown, A. (2015). *In time for tomorrow? The carbon conversations handbook*. Stirling, Scotland: Surefoot effect.

Richards, E. C. (2017). *The carbon literacy: Knowledge (CLK) E-learning effectiveness report*. Masters dissertation online at: http://www.carbonliteracy. com/research/. Accessed 12 June 2018.

Rucchetto, A., & Poland, B. (2015). *Lessons from inner transition in Canada*. Transition Network guest blog [online]. Available from: https://transition-network.org/news-and-blog/guest-blog-by-anne-rucchetto-and-blake-poland-lessons-from-inner-transition-in-canada/. Accessed 14 June 2017.

Rust, M. J. (2008). Climate on the couch. *Psychotherapy and Politics International, 6*(3), 157–170.

Seed, J., Macy, J., Fleming, P., & Naess, A. (1988). *Thinking like a mountain: Towards a council of all beings.* Gabriola Island, BC, Canada: New Catalyst Books.

Seeds for Change: https://seedsforchange.org.uk/. Accessed 2 June 2018 [Online].

Stock Whitaker, D. (2001). *Using groups to help people* (2nd ed.). Hove, UK: Brunner-Routledge.

Stoll-Kleemann, S., O'Riordan, T., & Jaeger, C. (2001). The psychology of denial concerning climate mitigation measures: Evidence from Swiss focus groups. *Global Environmental Change, 11,* 107–117.

Tripod Training: https://tripodtraining.org/. Training, facilitation and mediation for groups working on social justice. Accessed 2 June 2018 [Online].

Weintrobe, S. (Ed.). (2013). *Engaging with climate change: Psychoanalytic and interdisciplinary perspectives.* London: Routledge.

Weller, F. (2015). *The wild edge of sorrow: Rituals of renewal and the sacred work of grief.* Berkeley, CA: North Atlantic Books.

Wenger, E. (1998). *Communities of practice: Learning, meaning, and identity.* Cambridge: Cambridge University Press.

Whitehead, M., Jones, R., Lilley, R., Pykett, J., Howell, R. (2017). *Neuroliberalism: Behavioural government in the twenty-first century.* Abingdon: Routledge.

9

Leading with Nature in Mind

Rembrandt Zegers

Introduction

Nature (non-human animals and the material world) has been thought for a very long time to be without agency, contrary to humans who know (and express) values, have emotions, needs, interests and a will to reflect on and direct actions from. That idea is changing, and nature now more and more is seen as having agency and being an actor in its own right (Haraway 2008; Despret 2013; Bennett 2010). This means nothing less than saying that the relation between humans and nature is actually dynamic (and not static as historically it was thought to be). This dynamic is based on exchange of information and reciprocity of influence. Nature metaphorically speaking 'talks back' (Cianchi 2015) but not through human language. So, if humans and nature do not share human language, then how can we think of human/nature relatedness?

R. Zegers (✉)
Voorburg, The Netherlands
e-mail: Re.zegers@planet.nl

© The Author(s) 2019
P. Hoggett (ed.), *Climate Psychology*, Studies in the Psychosocial,
https://doi.org/10.1007/978-3-030-11741-2_9

One of the philosophers that has influenced the understanding of relatedness of humans and nature is Maurice Merleau-Ponty. He was one of several who responded to Descartes who had split the mind from the body in his theory of being human (cogito ergo sum). He considered the mind to be the 'machine' that makes sense of the world by taking in projections of it. In doing so he introduced a logic of exclusion, excluding as inferior anything that was not knowledgeable in the same way as humans. Descartes provided the vision and focus that became the foundation of (western) society. In contrast Merleau-Ponty investigated perception, eventually arguing that knowledge comes from the body rather than the mind, from the senses that interrelate with the world (Raghuramaju 2016). Toadvine explains:

> Sensing, in contrast with knowing, is a "living communication with the world that makes it present to us as the familiar place of our life"[1] investing the perceived world with meanings and values that refer essentially to our bodies and lives. (Toadvine 2016)

To give Merleau-Ponty's famous example, when your left hand touches your right hand each can be brought into focus, like a transparent cube that one can see differently from above or from underneath. This is like the left and right hand negotiating meaning. The same thing happens between humans and nature, as both 'sense'. Consciousness comes from this interaction, the middle ground between oneself and somebody or something else. That means perception is the start of knowing (Liberman 2007).

Toadvine insists that:

> the articulation of a new environmental ethic would be superfluous if we could not understand the world differently. (Toadvine 2009, p. 134)

This different understanding comes from closely looking at how the senses work in the world in creating meaning, for instance the meaning

---

[1]Merleau-Ponty (1962, p. 53).

of 'oneself'. In exchanging a gaze with somebody else, the other sees me in a way that I will never be able to see myself and vice versa. In this way who I am as 'self' can be seen as a dialogue between senses (mine and the other person's). Toadvine showed that Merleau-Ponty's philosophy of perception was also a philosophy of nature (Toadvine 2009) because Merleau-Ponty replaces the 'environment' as something static and outside of us with something else, namely 'relating' (as we are part of nature) in the form of a kind of dialogue. A dialogue that is unfolding and continuous. In place of the Cartesian 'knowing' about nature we become an instrument through which nature expresses itself. While at the same time humans have the capacity to reflect and create meaning from experience, closely sensing 'the world' as 'the world' senses itself through us.

Nature is usually not described or addressed in 'relational terms' (we call it 'environment' and in doing so detach ourselves from it; see and use it as resource that we can manage). But reality is much more dynamic as we are part of 'the thing' of the phenomena of nature and should not look at it in isolation. This is what we notice if we study perception. Through his theory of perception Merleau-Ponty overthrows the duality that comes with Cartesian philosophy, the duality and discontinuity of the physical and the mental, the objective and the subjective.

This makes Merleau-Ponty's philosophy an intersubjective philosophy where the creation of meaning (or to put it differently 'what is at stake') occurs not only between humans, but between humans and non-humans /nature as well. So, then nature becomes the very process of relating, the very process of exchange and joined 'meaning making' based on awareness of the intersubjective character of bodies engaging the senses. The Cartesian notion of humans 'thinking up the world' can no longer be seen as the best way to look at the relation with nature. That is, we are not manufactured by an isolated brain. According to Merleau-Ponty our brain is very passive in this process, more like a note taker (Liberman 2007).

From this position, the implication for leading (ourselves, others, practices) is tremendous, as it shows that cognitive knowing without reference to experience of how we are part of something can only produce what is already known. It also shows that leadership is a relational activity. While in general we do accept other humans as having agency alongside ourselves, nature needs to be added to that list of agents.

Leaving the philosophical arguments now for what they are, I want to make a few comments on the method of research before describing the three cases. To research the human relation to nature, methods that allow us to look at lived experience seem to be needed. This is so because our perception is largely unconscious. In this way nature never lets itself be fully known, but can be 'caught' (or interpreted) through carefully reporting and accounting stories or make observations of 'practices'. But doing that in a way that it is grounded in a research tradition of researching the human unconscious. One way of doing that is through using interview techniques that allow for the interviewee to tell biographical stories (Hollway and Jefferson 2013).

Phenomenology is similar in that its methods are also suitable for researching lived experience. They emphasize that element of staying out of 'what is already safely familiar' as Hollway and Jefferson put it. Finlay in her overview of doing analysis from a phenomenological perspective states: 'Thus, theory, explanation, judgments, and the researcher's previous experience and beliefs are temporarily pushed aside to probe the "is-ness" of the phenomenon further. Critical attention is paid to how the phenomenon is presenting, specifically how it is experienced by participants in their natural attitude. Researchers accept that what participants say about their own experience is their "truth" and do not morally judge. Their focus is on the meaning of the situation as it is given in the participant's experience' (Finlay 2014, p. 123).

Following now are three cases, based on original interviews by the author.

Case 1: The Girl Who Loved Horses and Now Has Them

Jane tells me she liked horses from when she was a little girl. A friend at school had a pony and Jane wanted one as well, being with a pony, *'grooming, riding, talking'* to it. Her parents allowed her riding lessons, but according to Jane could not imagine the concept of her owning a horse. She grew up in the countryside, cycling to school through the fields. In early adolescence the family disintegrated, she got ill and then needed to focus on finishing her school, all reasons she tells me why horse riding got lost. Her mother is mentioned several times in the story. Where her mother liked flora, Jane liked fauna. Her younger brother brought animals home. Her mother was influential in politics. Jane tells me she felt her mother tried to save her children from nuclear weapons, among other things such as trying to keep the national airport as small as possible. Jane tells me that in that period of her life she could sense *'feelings of destruction'*. *'If people make war against nature, it will eventually cost us our lives'*, is what she tells me she heard from her mother.

Jane states she felt more at ease in nature than at home. When at home she would prefer to do her home work in the attic. If playing she preferred to do that out of sight of her parents, with Barbie dolls and horses. She describes her parents as not giving enough guidance, as avoiding talking about the issues that needed to be talked about. She felt unsafe. When she visited friends their parents seemed much more helpful, supportive and focused on their children.

At first the NGO where she worked gave her satisfaction, being in the frontline, experiencing camaraderie, excited about being so mission driven together with colleagues. After a while when being more in management roles, Jane lost her joy. She explains how now, years later, she sees the culture of NGO's attracting a certain kind of person who strongly looks for likeminded others. It is also a culture which does not pay much attention to people's emotional well-being.

In those later years in the NGO in management roles, Jane wanted to know more about herself and why she was so highly strung. She started to realize that most of the time she was rather stressed. She worked with a therapist and started to pick up horse riding again.

There was an exercise that made it very clear to me that actually I had come that far in life with unconscious trauma. In that exercise, I was asked to retrieve a horse from the stable and in doing so I had to indicate the level of my comfort on a scale from zero to ten. I was expecting to be zero, as taking a horse from the stable is what I did every day. But when I focused on answering the question I had to answer that I felt like a five.

That is when I realized my problem, being so used to stress all the time that I didn't even recognize it. I learned that dissociation is what people with post-traumatic stress syndrome fear most. So, a five is actually on the way to tense and evidence of constant fear. It struck me that I had actually lived my whole life on adrenaline, with a lot of tension. That day brought that insight to me. Realizing this has given me back my life, being able to understand how my life had been, as it had been. Why I could not achieve certain goals because you are actually busy processing trauma all the time.

Following on this period in her life she took up a specific training in America to work with horses and people. Now she has her own company working with management teams helping them developing their work relations.

On the way back in the car I still asked some questions. What is she learning herself at the moment, how does she see her development? She has not abandoned all management practice for herself and her business; as she states, she not only uses her experience in large organizations as a way to understand what her clients are experiencing and in what managerial jobs they spend most of their time, but she is also now looking for sound business advice, as she wants to create a successful marketing strategy for a next phase of her business. But although she intellectually knows what she could do, she is finding it difficult to develop concrete activities from the things she academically already knows. She thinks she needs to develop a routine process for the marketing of her business.

Jane offered to work with me so that I could have an experience of being with a horse. The following is a description of my own experience in working with Jane and a horse. (I use the present tense in the following.)

Jane first asks me to approach the horse. I do. I walk towards it from about 10–15 meters distance. It doesn't move away from me. Jane explains she sees the horse accepting me as a person. I talk to the horse and expect it to look up or something (like with my dog). But it doesn't look at me, in fact it doesn't seem to respond at all. I discuss this with Jane, she tells me that the horse, being a prey animal, checks me for safety and probably not for the content of my kind whispers.

Jane considers my need to establish an exchange at this moment with this horse through language, whispering my good intentions to it, as a projection of wanting to be loved, to be seen as nice. I find that shocking, but revealing. She explains that such learning is what horses provide, being projection screens. Jane asks me what I want to do next.

The horse lies down. I have an idea, an urge to lie with it, not necessarily touching, curious how that will be and if I can stay relaxed. I think I want to feel like a horse. I lie down and after a while feel my body and I start wondering what I want to do next. I want to stand up again. I feel I want the horse to stand up as well. The 'want it to stand up feeling' is coming out of my body, I do not think about it. However, I feel determined that the horse should move and before I know it, it stands up.

I want to lead it and we start walking, me and the horse. I sense a connection, a presence, I am absolutely not alone and aware of encountering the energy of somebody else.

To my surprise I am able to make the horse move within a relatively short time (I am proud). I sense how to do it (with some suggestions from Jane) by using my body. I am no longer talking to the horse, I am leading it by starting to run from a distance from behind the horse. I am finding out that it is not about banging into the horse, but actually more like almost passing the horse in a parallel movement but still making sure I keep behind it at an angle. I can now even start to use the outer fence of the exercise ring so that the horse is in between the fence and myself, and I can make it walk in a circle around the circumference of the ring, or let it make a short cut when giving it more space to turn. I am getting the feel for the distance I have to keep from the horse, the pace and the right angle to keep it moving.

When talking about this exercise I started to find it plausible that the horse was responding to my energy and me in turn to the horse's energy as I felt clearly in the lead but in an interaction as well, feeling it's response. It is hardly possible to explain on paper, but my experience was as if being with the horse, not against it. The horse was running with me, as I was running with the horse. Jane explained she was excited that I could do it and said it had to do with allowing myself to 'go out of my head', not 'thinking' or 'reasoning' as in human conversation.

I felt I had learned something about myself and horses and about the work Jane does.

Case 2: 'My Friend and I Became Herders'

Daniel decided to do herding on a remote mountain together with a friend, when he was about 20 years old, still living in the city where he was born. It was a period just after the 1960s when young people in cities were looking and practicing newly acquired freedom in society and also becoming aware of the environment and its concerns. Daniel's older brother had been active on the streets during student revolts.

Daniel tells me about the farmer communities and the hierarchy, with the community president keeping an eye on the herders that work for the collective. Herders are lowest in rank, at the same time very important doing the responsible work of looking after cows (so the cows come back in one piece from the mountain) and producing cheese with high economic value. Not all mountainous areas are managed as commons but a lot still are.

As Daniel became more experienced, he started to read the cows in relation to the land, the weather, his colleagues, the dogs and the farmers (as the latter sometimes visit unexpectedly). For example, he saw that the cows are restless at the beginning of the season, happy to be out again, most of them expecting calves. Then giving lots of milk at the beginning of the season, but later it gets less. So does the quality of the grass as it changes towards the end of the season. Cows can suddenly move to places between the trees to protect themselves from the weather—then you don't see them any more—or by accident they can get stuck between rocks—then you have to decide if you need help from the farmers. Young cows can suddenly become 'crazy', starting to play uncontrollably. Older cows can have disputes and not 'speak' to each other anymore. Rivers can suddenly become floods that you cannot pass.

Daniel tells me of a time, during the 5th or 6th year of his herding life, when he and his first wife—who had started to join the summer herding job— were on a new mountain area. To his surprise one day, as he had been herding in the area for a while, a local farmer told him about a plan to build a dam and lake in the area. Daniel was in disbelief. The next season he noticed people. They were actually making small explosions in order to find out about the composition and strength of the rock. In his story Daniel gave a lively account of his protesting and wanting to see the bosses. He felt they had no right to destroy this beautiful area, 'his' area as he wholeheartedly felt it. He loved so much this land and knew every corner and twist of it. He explained in the interview how he was not sure at first but then felt he had to take up and organize the locals to not let this dam be built. But it was a long journey of several years before the mountain was eventually 'saved'. As Daniel told me many details, what he emphasized was the reluctance of the farmers and local villagers to protest, even if they did not support the build. They felt it was not possible to oppose the authorities on this, they were dealing with a very powerful electricity board, operating in large parts of the country, that also supported communities directly in many ways, sponsoring football teams etc. Although a fair majority of the locals were not interested in the dam itself or even opposed to it being built, Daniel experienced how difficult it was to involve local people, neither did they want to confront other locals, basically they didn't want to make any trouble.

Daniel uses the expression 'being in the middle'. He described and meant it I think in two ways. One, the way that in ordinary conversation people can say they are 'in the middle' of doing something or 'in the middle' of a situation or dispute with other people. But Daniel also meant it in a different way (as did one of the other people I interviewed in a different place), namely in the middle or in between people and nature, as if being in separate worlds at the same time, or switching, or at least being aware of worlds with different ways of being. Apparently, the experience of being on the mountain in nature and the experience of being in a more inhabited place (a farm or the village) were felt to be of such different character that one could feel in the middle, in between. I associated it with translation, with being between languages or cultures. I also associated it with a very private experience, one that cannot be easily explained or shared. An experience that one has to find words for.

Doing his herding work in the mountains, being in the open and most of the time with just one or two other persons, he stated that coming back to the mountains in the following season always felt as coming home.

> After three months, you go to the city. After my first mountain seasons, it was crazy. I went to the city, you sit in your VW bus with all your cartons (RZ of belongings), in it you drive to your I don't know where house, which you organized for winter because you didn't earn so much money, so you had no permanent home, because you try not to pay the rent during summer, so you arrange something new. You drive there and you realize you look into every body's face much too long and you read every advertisement on every window of a bus or on a wall of a house you read it and you think what does 'Excellent Insurance' mean to me. Until you realize, I have to stop reading every letter. F... why did I read every letter. Well because I didn't read so much when on the mountain and I like reading. And I am probably now attracted too much to any new sentence. Before, any face that showed up was important because it was a farmer. Sometimes you also met a hiker and then it was quite interesting to have half an hour talk with a hiker, so that was an important face in that day. Now suddenly there are a hundred thousand people because it is rush hour in this driving through city. Daniel, I tell myself, stop watching every face. Makes no sense. It is another world, I need other filters, I need other behaviors. You know you can do it. But it has to happen fast or you will become crazy.

After his herding career Daniel has been a director of different organizations. He tells me he knows what sustainability really means. He also tells me how he experienced nature and the animals he worked with as having agency as actors. Most interesting is his comment that working in the mountains with animals has influenced the way he is a leader of organizations:

> Be present, have the overview. Don't do anything as long as they know what they do. Make your plans for the next few weeks, years, months to come, whatever is needed. Be close to them, if they have a problem think about if you really can help, if you have to help and can help go and do

the job and then let them do their job again. Until I had, until I had got that really as a herdsman. It took me years. It took me years. Especially because I was not trained in cattle or cattle behavior because of growing up in a big city. I was 'over managing' for 6 or 7 years. Until I was placed in this huge mountain area, as the smaller the pastures are the less you learn as a herdsman. The bigger the size of your mountain area, just as the bigger your company is, the more you have to be aware of not interfering, but you have to know what is going on in that herd.

Case 3: 'When I Was Fishing One Day, I Had This Idea'

Gunther and I spoke in a café near the railway station half way between the village where he lives and works as a director and teacher in a school and his other job at University in a large city, where he goes once in a while. He was born high up in the mountains. He explains to me that he loves snow and that the colors of blue sky and reflecting light on snow affect him deeply. He wanted to become a skiing teacher and did everything to be trained as one, but then realized he would never excel at it and decided to study geography and learn about glaciers especially. He comments that the motivation for that came through dreaming and this motivation became key in working through his studies and succeeding in doing a Ph.D. on the topic of glaciers. He knows a lot about them and he is involved still in the science of glaciers working with some of the world's well-known experts.

Almost immediately in the interview he tells me about making music on every occasion when he and his international research partner lecture on the topic of glaciers and global warming. He was first invited by this friend to make music and he learned to play in the style his friend suggested. He has started two orchestras now, both are associated with glaciers. He tells me that speaking about what is going wrong in relation to nature doesn't lead to action and music helps to motivate people as it offers a different space for relating to the topic of the melting glaciers. He also mentions the scientific studies on motivation done by a colleague of his at university that show that motivation is the most important element for social change. This relates directly to his own life and his dream to work on glaciers. One story was about how he played music during a lecture where a young man attended and that he was called some days later by this man. He wanted Gunther to come and do the same kind of lecturing, making music for the association of mountain guides. He tells me this as an example of how music works in motivating people, as people can be emotionally touched as humans with potential and not only be overwhelmed by these terrible alarming messages. He shows me pictures which provide evidence of the speed with which some of the largest glaciers are diminishing in size. Then he tells me another story.

> Then the next day I was fishing and contemplating on the fact that every day now for some time it had not rained and thinking what the 'rain situation' meant for the amount of melting water from the glaciers. Then I remembered the phrase from Voltaire, 'we are also responsible for the things we are not doing'. This has everything to do with motivation, the ice music stories. Maybe in 20 to 30 years the glaciers have no fresh water storage capacity any more. Then I had the idea that we could do melt water recycling, maybe using the invention by the Ladakh people who build wooden ice towers over winter, releasing their water in spring and early summer, as Ladakh has the same problem with melting glaciers. I told this to the CEO then he said 'do you really think so'. I said 'Maybe it could'. 'Write to the authorities' he said. My first step was to phone my colleague abroad. He said, 'I have to think about it'. After some days, he came back telling me he thought it could work. Then I wrote the letter to the authorities. They were surprised. They gave some money to analyse if it is possible. It is clear that this has become important to investigate for the future of freshwater storage.

Valuing Nature

Each person in these cases values nature in a unique way. Jane tells about how working with horses has helped her psychologically, Daniel wants to protect a place which is close to his heart and prevent a power company from transforming it into a lake: a place that he came to love for the life he can live there and the tradition of making use of it through herding (itself a rather sustainable practice). Gunther cares deeply about glaciers with a view of the future. He maybe thinks more of the function that nature has for the community, providing resources. The similarity in all three cases is that there is a personal resonance, a personal story where nature and self cooperate.

In the stories, the decisions to follow through a path of action are not explained in terms of calculated planning, although some planning is needed in executing decisions. It is as if the decisions come from insights, moments of clarity about who one thinks one is and what one should do in a particular context, where nature plays an important role. It seems to me in knowing what to do, particular forms of going forward unfold, that are not questioned and at the same time are unknown. Yet there is a great deal of confidence in each of these cases. Their meaning making has its origin in the experience of nature and self, as there is no hierarchy or previously held career plan involved. There is a vision of what should be, what is right and desirable, that not only has to do with self, but also with the collective, the commons and with nature and how that connection can be kept.

In psychoanalysis, the unconscious is what is called a representational unconscious. In the most basic idea of therapy certain emotions in the history of a person have not been integrated and have influenced that person's mental development. In psychoanalysis field theory now considers the non-human environment as also important (Ferro and Civitarese 2015). As perception of the world is meant to make the phenomena in the world conscious, at the same time the senses work largely at a preconscious level (e.g. most external stimuli are filtered). Unselfconscious influences from nature on psychic functioning in this way should be considered important as such, being influences from a person's social history and upbringing. The reality of the senses helping conscious perception as well as operating on an unconscious level can be found for instance in Daniel's story. It is remarkable how he talks about the transition he has to make when coming back from the mountains to work again over the winter in the city. He talks about needing to close off his senses when returning back to the city. He also talks about a feeling of home coming, or being at home when in the mountains.

In these case examples, the very element of working with nature shows itself to be dynamic. But this can require hard work learning to relate differently from what one is used to. For instance, Daniel had to learn to 'listen' to the animals, he had to learn how to work with the cows and the dogs. That was not easy for him. It was only possible when he reached a crisis, not knowing what to do or think any more to handle the situation. He had to relearn problem solving, through a different way of listening. He had to learn to observe the cows, read their behavior in the landscape. Equally he had to listen to the farmers and learn from them. Stopping what he was doing and accepting the knowledge the farmers had to offer, even when at first, he had no intention of listening to the famers. Although language helped obviously in communicating with the farmers, learning to communicate with the cows and the dogs was different. This leads to another insight; the case stories reveal three people on the edge of being with or in nature and in a social human 'normative' reality at the same time. For instance, Gunther is expected to consult with his superiors about his idea of freshwater conservation, he knows he has to do that as part of the way 'things go' if he wants to be successful and create change. He has to make his idea fit with social conventions, including the way science is involved and needs to be applied to make the idea feasible. He needs to work the established managerial and institutional relations. He has to engage with those who have social status and power in the community for them to know about this idea and maybe buy into it. Jane knew it can be helpful for her to know about marketing and how you do that, following the current modern knowledge and practices of 'selling your work'.

It was Harold Searles who saw the role of the external non-human environment in psychic development and functioning when he was therapeutically working with people suffering from severe mental illness. He started to include the relation to nature in psychoanalytic thinking and theory, stating that the relation was of vital importance. In his 1962 book 'The Nonhuman Environment in Normal Development and in Schizophrenia' he elaborates on human development as an individuation process where one learns (differentiates oneself) to be able to see oneself as separate (distinct) from other humans. Searles shows this (separating oneself) holds true for the non-human environment as well. He illustrates this with many examples of people suffering schizophrenia that tell about being one with (an element of) the non-human environment/ nature, like a stone or part of a house. Such a stone, for instance, then talks through the person and that is considered crazy, or in clinical terms a reversed differentiation, as young children can feel themselves as one not only with people (mother) but with an animal or nature as well (Searles 1960).

Some of the experiences of being able to communicate with animals (or sensing nature in a wider context) as described in the cases here, could be interpreted either way, as expressions of sensitive but mentally healthy people or as expressions of 'crazy people'. The people interviewed for this study can safely be seen as not to be suffering from schizophrenia.

But something else is important to note, namely that these three are very aware of, and cautious about, the extent they can or should share their experiences of nature. These cases show how the people assert themselves and how comfortable they are with their own meaning making when it comes to nature and their creativity in working with nature. What is experiencing nature 'worth' individually and socially, if it cannot openly be supported or explained through (scientific) knowledge or in shared reflective knowing? Several interviewees explained they felt they could not share or only (much) later in their lives, as they were not able to find the people they could share their experience with, or they were afraid of not being heard, or being seen as weird or that telling about their experiences could be controversial. When Daniel states he feels in 'the middle', I take that as a comment about how he experiences culture and nature being split.

As eco psychology and more recently eco sociology have developed, the relation to nature and the ecological self as the self that is connected to nature are becoming key concepts. Equally, the interest in relating to nature is driven by research into health issues and the overwhelming evidence that even the smallest exposure to nature (like plants in an office, or even nature pictures on a wall) can have an impact on physiological health markers like blood pressure etc. (Frumkin et al. 2017).

Searles drew attention to the vital importance of healthy mental relations towards nature (Searles 1960). At the same time Searles pointed at the natural sciences as the main players to add to knowledge about the natural world and the challenge for modern humankind to integrate that knowledge to become a better kind. But in many ways the opposite has happened. Cartesian philosophy has helped western humans to create a culture of reflective knowing, but one that is seriously (if not dangerously) 'detached' from nature. Merleau-Ponty's philosophy of perception builds on the notion of everything interconnected with everything else, opening up to perception. This in turn can help us to experience nature as having agency. This is what I took from the three cases and certainly also from my own experience becoming aware of how I could actively and dynamically relate to a horse. Potentially that could recreate our relationship to nature from a more eco-centric perspective (Toadvine 2009).

References

Bennett, J. (2010). *Vibrant matter, a political ecology of things*. Durham and London: Duke University Press.

Cianchi, J. (2015). *Radical environmentalism, nature, identity and more-than-human agency*. Basingstoke: Palgrave Macmillan.

Despret, V. (2013). From secret agents to interagency. *History and Theory, Theme Issue, 52,* 29–44.

Ferro, A., & Civitarese, G. (2015). *The analytic field and its transformations*. London: Karnac.

Finlay, L. (2014). Engaging phenomenological analysis. *Qualitative Research in Psychology, 11*(2), 121–141.

Frumkin, H., et al. (2017). Nature contact and human health: A research agenda. *Environmental Health Perspectives, 125*(7), 075001. https://doi.org/10.1289/EHP1663.

Haraway, D. (2008). Training in the contact zone: Power, play, and invention in the sport of agility. In B. Da Costa & K. Philip (Eds.), *Tactical biopolitics* (pp. 445–464). Casmbridge, MA: MIT Press.

Hollway, W., & Jefferson, T. (2000, 2013). *Doing qualitative research differently*. London: Sage.

Liberman, K. (2007). *An inquiry into the intercorporeal relations between humans and the earth*. In S. L. Cataldi & W. S. Hamrick (Eds.), *Merleau-Ponty and environmental philosophy, dwelling on the landscapes of thought* (pp. 37–49). Albany: State University New York Press.

Merleau-Ponty, M. (1962, reprinted 2002). *Phenomenology of perception.* London: Routledge.

Searles, H. (1960). *The nonhuman environment in normal development and in schizophrenia.* New York: International Universities Press.

Toadvine, T. (2009). *Merleau-Ponty's philosophy of nature.* Evanston, IL: Northwestern University Press.

Toadvine, T. (2016). Maurice Merleau-Ponty. In Edward N. Zalta (Eds.), *The Stanford encyclopedia of philosophy* (Autumn 2016 Edition). https://plato. stanford.edu/cite.html.

Online Video

Raghuramaju, A. (2016, February 17). *Merleau-Ponty's phenomenological epistemology* (Online Video). Available from: https://www.youtube.com/ watch?v=NpyX2ho_1Gs. Accessed 22 Feb 2017.

10

Attitudes to Climate Change in Some English Local Authorities: Varying Sense of Agency in Denial and Hope

Gill Westcott

UK local authorities have in recent years been charged with part of the responsibility for reducing climate impacts. The research reported here began in 2010, as the Government's Climate Change Act had prompted the setting of new goals and incentives for councils to reduce carbon emissions from their own activities and to encourage similar activity within their communities and businesses. At the same time the financial crisis of 2008–2009 and the ensuing austerity was greatly restricting central funding for local authorities' budgets.

This research investigated attitudes to climate change among local authority personnel, whether and how these influenced the actions that authorities have taken, and how collectively determined norms and procedures interlinked with individual attitudes to affect the outcomes.

G. Westcott (✉)
Venbridge House, Cheriton Bishop, Exeter, UK

© The Author(s) 2019   **195**
P. Hoggett (ed.), *Climate Psychology*, Studies in the Psychosocial,
https://doi.org/10.1007/978-3-030-11741-2_10

Climate Change: Response and Denial

Despite the recognized presence of literal denial of climate change or of its human causation, studies by Stoll-Kleemann et al. (2001) and Norgaard (2006) examine the more common failure to recognize the implications, while acknowledging the existence, of climate change and to explore possible courses of individual or collective action.

Failure to acknowledge anthropogenic causation would be termed 'interpretive denial' in Stanley Cohen's classification derived from studying denial in relation to atrocities and suffering (Cohen 2001). Failure to consider remedies would be 'implicatory denial', also labelled 'stealth denial' by Rowson (2013). Weintrobe, too, distinguishes complete denial that something is so, from 'disavowal', a partial recognition but one which denies the significance, magnitude and painfulness of a circumstance or issue. The latter, she suggests, is likely to be more lasting than literal denial, less falsifiable and 'artful': reality can be 'seen and not seen at the same time' (Weintrobe 2013). The present study therefore sought to distinguish this partial recognition from the literal denial of climate change. The term climate 'dismissal' may be used to avoid prejudging which theories most adequately account for the observed lack of attention given to the risks which climate change poses.

Methods

Through semi-structured interviews in 2011–2013, the author spoke to 34 district and county councillors and council officers in the South West of England. A questionnaire was used to elicit information about actions and achievements. Brief statements about climate change from a variety of perspectives were then presented on cards, inviting respondents to reflect more freely on their own views and how this affected their actions and policy outlook (Westcott 2017). To understand attitudes more fully, the resulting narratives were analysed using the psychosocial methods described in this collection. This was particularly helpful in exploring organizational culture, norms and collectively policed limits to discourse, as pioneered by Norgaard (2011). Where something is ignored or denied, analysis of face value statements falls short in identifying what is being defended against, whereas a psycho-analytically informed reading of the narrative can reveal the process of denial and the role of particular stories or implied meanings in avoiding painful or socially unacceptable emotions, as illustrated below.

Interviews were conducted with councillors and council officers whose responsibilities were relevant to carbon emissions, with additional councillors in two contrasting authorities, intended to deepen understanding of the local political context in which central government carbon reduction policy was being carried out. The case study locations were Greenleigh, a sparsely populated rural area with pretty coastal settlements attractive to incomers and high topography suited to wind energy; and Weirbridge, a county town with good transport connections, new industrial estates and a fertile hinterland.

The interviews with officers and members of these and five other local authorities in South West England, combined with statistical data, show that considerable reductions in councils' own greenhouse emissions have been achieved, not wholly due to cutbacks and other contextual factors, but measures to achieve more widespread carbon reduction in their geographical areas were more limited. Central government finance and policy were key determinants of action in all the local authorities studied, but significant differences in the approach of different authorities remained.

Views and Attitudes About Climate Change

Table 8.1 classifies the respondents' expressed views about climate change. The sample is not representative of councillors and officers in these authorities, but includes disproportionate numbers of those whose interests or training had taken them into environmental health, energy management or development management.

A large majority of respondents did view climate change as a problem which was immediate and urgent. Yet, despite the presumed pro-environmental bias in the sample, seven individuals (25% of those who commented on this) denied or doubted the human causation of climate change, and 6 (21%) viewed climate change as a problem much less urgent than others, perhaps in the 'nice to have' category, like the arts. In the general population of councillors and local authority staff the proportion of those who dismiss climate change as a problem may be considerably greater.

Views about climate change were also rather vague. Just under half could offer any comment or opinion on statements concerning the speed or scale of likely climatic changes, or on the possibility of tipping points in climate systems. Two of those who did comment rejected the view that there are tipping points. Many respondents had never heard that environmental refugees are already leaving Pacific islands, delta settlements in Bangladesh, and areas undergoing desertification in Africa, the Middle East and China. However for impacts close at hand, knowledge was more precise. For coastal communities, sea level rise was very much on the agenda, and remedial action for flooding in Weirbridge was the subject of discussions with the Environment Agency. The implications of higher precipitation for road maintenance are also taken into account in a matter-of-fact way by Highways Departments (perhaps because this reinforces their claim on resources—but this is to anticipate!).

Urgency and Action

Many of the officers I spoke to were selected for interview by virtue of their responsibility for energy management, recycling or environmental health. Often they spoke of climate change as a key concern and motivator for their activities. Bernard, an energy manager in a mixed urban/rural district, described the balance of views in his council:

> I think there's a strong faction among the elected member who believe [that climate change is mainly a natural occurrence]. But there's also a passionate group of members who don't subscribe to that at all. Dare I say it, I would say that senior officers here believe in climate change.

In contrast to many other respondents, with the support of his director and a senior councillor, Bernard was able actively to seek energy efficiencies and funds for low carbon investment.

Alison, a director, had no hesitations rejecting the text of a stimulus card referring to climate science being of doubtful value:

> Right I can discard that one immediately…….. (rapidly) I don't think it's natural. (thump as she puts the card on the table) Don't think it's a hoax. (thump)(thump)(thump)

Gerard, a senior county councillor who has held environmental health portfolios, added a moral flavour:

> of course there have been natural occurrences that have caused episodes of warming and cooling of the planet, but this is the first time that it's been activated by man's greed and avarice.

He spoke of the likely cynicism of young people when they see that little is done about climate change, later adding

> But there is that overriding sense of, OK, yea, but what can I do about it? It's becoming aware, but then lifestyle changes have to be made. You know, I've got to use my car less [followed by review of other lifestyle measures].

John, an Independent councillor in Greenleigh, also thought the issue was urgent, but was pessimistic about concerted action:

> You're going to burn that coal. There's that, you know. Your immediate needs…….

> If I said to my constituents…, when it comes to elections, I'm going to [support] wind turbines all along the villages, it's your environment being sensitive to your grandchildren in years to come, I know that I wouldn't get elected again…

Sceptical Views

Among those regarding climate science as suspect was Dorothy, a councillor in an authority which has been very proactive on the carbon agenda.

I studied physical geography at University, including climates and this sort of thing (chuckles) and the climate's been changing for millennia… Yes, it's changing now quite rapidly, but it's changed just as rapidly in the past, before man's influence. As far as I'm concerned the jury is still out on what man's involvement in that is.

One respondent reported a councillor objecting to mitigation measures because climate change is occurring naturally and manifests God's will. Richard, a manager of council property, told me:

You will get totally different opinions on it, you know. I don't think people are convinced about climate change and its causes ………and there's that distrust of scientists, …. (laughs)…I mean, ….they're so definitive ….and you think, well, before noon, they were saying everything was black and white, and afternoon came round and they said white and black, …..

In Greenleigh I was told firmly by Alan, a councillor

I think we don't know enough about climate. We know there are wide natural variations, between ice ages, and we don't know what part we're in. And we cannot at the moment graph out what would happen on a daily, weekly, monthly or annual basis, for the last 200,000 years even, let alone for the last 2 million years, which would probably be a better example. …. We hear a lot about global warming, and yet, everything is pretty much colder really. …. I think there are natural cycles. It's very dangerous to blame it on human activity.

Alan was not the only one to question whether warming was in fact happening. Jack, Leader of Greenleigh council, commented '*you tell me about global warming, a month ago, when we were at minus ten, minus twelve*'. Such statements were reported in council chambers when urging refusal of planning permission for wind turbines.

Kate, a councillor from what was perceived as a 'rough' housing estate, was the most fatalistic in her attitude to climate change. A woman of enormous energy, she had founded a Food Bank for households unable to afford to feed themselves, and responded frequently to housing and benefit problems. She did not deny that climate change was bad news, but was clear that '*It's a natural cycle. It's going to happen no matter what. I do believe that. ...... Humans may try to fight it but they're not going to stop it*'. Statements like this could be taken as evidence of ignorance of science or denial of human agency. But in context (Kate had also retired from the planning committee, finding it unresponsive to her opinions and preferring to 'fight from the outside'), it seems possible that her views reflect a projection of experiences of futility in the realm of human society and politics onto the realm of nature.

'Bottom of the Pile'

Several respondents thought that although climate change is largely anthropogenic, its importance has been overestimated. Even where councillors and council websites proclaim that carbon reduction is a high priority, the pattern of spending and the discourse about budgets tends rather to indicate that carbon is in fact a rather low priority, if it figures at all.

Spending cuts, poverty and unemployment were the priorities, accompanied by half a dozen more headaches with much closer deadlines. The more senior the respondent, the more heavily responsibility for budgets weighed; and the short-term worries, including the prospect of having to carry the blame for cuts and redundancies, or being made redundant, seemed to crowd out longer term issues.

Jack had reported Parish Councillors' lowest priorities thus:

> Sadly 99 times out of 100, apart from Arts and Culture, it's usually anything to do with environment, climate change, it's at the bottom of the pile ... if you are a ... single mother feeding your family of three young children, your most important thing is that you clothe, feed and keep a home for that family. Far more important than working on sunshine or snow.

In fact, '*If you talk to most of the holiday makers who come down here in the summer, and if it went from 23 to 27 they'd be delighted cos they can go to the beach!*'

Jack distanced himself from the issue of climate change by referring to a possible timescale of 3000 years, but went on to consider the possibility of much more rapid changes. In fact, he almost exactly mirrors the divided state of opinion on the council on which he pulls together a minority administration. '*The left hand thinks it's a load of tosh and the right hand thinks it's the best thing since sliced bread, what do you do?*'... (followed by a rapid change of subject).

Beliefs or Strategies?

I have sought in Table 10.1 to classify the voiced opinions of respondents in line with the concepts of 'literal' (statement A), 'interpretive' (statements B and C) or 'implicatory denial' (statement D), as if the statements represented more or less settled belief. However, in a number of interviews it was hard to classify a person's views in this way because they expressed more than one view, sometimes one immediately after the other.

For example, Richard has what might be called a three line defence:

1. He doubts the findings of climate science: '*everybody always says things are black and white but they never are.*'

Table 10.1 Views about climate change

	Opinion	Number	Number who could comment on scientific statements
A	Climate change is not happening, or doubtful whether it is happening	0	
B	It is happening but is mainly a natural phenomenon	3	1 (disagreed there are tipping points)
C	The jury is still out on human causation, doubt about findings of climate science	4	2 (1 disagreed there are tipping points)
D	Climate change is anthropogenic but not urgent, or other problems are more pressing; or technology is changing more quickly than the climate	6	0
E	Climate change is urgent	15	8

2. Anyway the Chinese: '*There's no point in the West saving carbon emissions, if China and India are increasing by a bigger amount. What's the point?*'
3. He explains that council members will tend to veto major expenditure bids unless there is a strong business case to 'invest to save.'

The second two statements emphasize others' responsibility for the problem, in contrast to his own action, slipping in measures to reduce energy consumption in many other projects.

Again, a councillor who tells me that she is sure that climate change is mainly a result of natural causes, agrees not only that her council can play a part in mitigating climate change but that individuals can as well. And Charlotte, a senior councillor, says that climate science has really convinced her, but '*How are you going to do it if....you see, the worst people are China and America, or some Americans are aren't they?*'

When people seemed to hold differing opinions simultaneously, or, in one or two cases, were reluctant to own any of the various views they reported others as holding, it suggested to me that there might be a more fruitful way of looking at them. These statements are ways of constructing the issue, and perhaps they are best studied in context to see whether they serve a function; whether they tell a story which explains or gives meaning to the teller's position and practical response or lack of it to the issue of climate change.

Kari Marie Norgaard (2006, 2011) in her study of a small Norwegian town, draws on Cohen's typology, as I have done, but describes the positions as among a cultural repertoire of responses which people constantly use in everyday life, in order to manage unwelcome emotions such as fear, helplessness and guilt.

Norgaard draws on the work of social psychologists such as Zerubavel (2006), to explore how denial is socially organized. Avoiding certain topics or conclusions draws on cultural repertoires and norms regarding the control of attention and emotion, so that 'ignoring is done *in response to social circumstances*, and *is carried out through a process of social interaction*' (Norgaard 2011, p. 9, italics in the original). Accordingly, she identifies conversational gambits and statements people make as constituents of a cultural toolkit, a set of strategies and stories, to reconcile their awareness of the seriousness of climate change with their conformity to surrounding social norms and perspectives. The unwelcome awareness could evoke emotions and expressions which transgress the norms of small talk or social gatherings, and thus the 'cultural toolkit' strategies enable people to avoid attending to or thinking about climate change, particularly in social settings.

Norgaard (2011, p. 97) describes a variety of conversational and social strategies for selective attention, and the control of feelings, and therefore, of thoughts. These strategies are also used to control individuals who do not conform to conventional norms of expression and behaviour. In stressing that denial of all sorts of issues is a dimension of 'normal' behaviour—indeed is part of how 'normality' is constructed—Norgaard shows how not looking at the implications of climate change is deeply embedded in the warp and weft of social life. This perspective seems to give a more successful account of how inaction on climate change occurs on a social as well as a political level, and of why it may be so hard to change.

If scepticism or 'bottom of the pile' attitudes do represent a kind of strategy to justify inaction, then it is interesting that no such climate change dismissal occurred when I spoke with Grace, councillor from a small coastal town whose low lying centre is protected from storm surges only by the main railway line to Penzance which divides the shoreline and town centre. This line has been breached on several occasions and requires costly work to maintain against rising sea levels and higher storm surges linked with the increasing intensity of depressions from the Atlantic. '*Without Brunel we would not be there*', Grace told me. But Network Rail's commitment is not indefinite: '*They're giving us 20 years – no more!*'.

Repertoires, Norms and the Cultural Toolkit

Within the interviews, certain recurrent patterns or themes manifested attitudes shared across members of an authority, political party or society at large.

The Business Case

Although in circumstances of financial austerity, there is ample justification for the business case to hold a prominent place in councils' decision, the near universal dominance and acceptability of an economic perspective seems to be a social phenomenon which forms a unifying discourse, accepted by all parties with differing ideologies and worldviews. Indeed, at national level, reference to whether policies have successfully promoted economic growth is often an unquestioned and sometimes supreme criterion of value.

However, reference to the business case is by no means universal, for example in local authority decisions involving cycle tracks, pedestrian safety, street lighting, and child safety. Yet decisions affecting carbon emissions appear seldom to benefit from evaluation in terms of social or environmental impact which presumably is more contentious.

Ignoring, Minimizing and the Application of Flippancy

Council officers on the whole attended to the questions I asked and answered them succinctly; the elected members' interviews typically took an hour to two hours, sometimes longer, yet I was surprised that instances of difficult questions being ignored would arise so readily. For example, when I was talking with Charlotte, a very experienced councillor, about her perspective on the future, this exchange occurred:

> G: So taking it all in all - these are quite long term questions some of them – I wonder how you feel about the future. I don't know whether you have grandchildren?
> C: I have six – two in New Zealand and four in Gloucestershire.
> G: So a worldwide family.
> C: No, of course I do. One thinks of their….but I tend to be realistic. How realistic is it – I don't know - you can only go with your common sense. I hope I've got some common sense. What are your views on nuclear then?

Perhaps this councillor's reply illustrates the repression of a degree of anxiety (provoked by my reference to grandchildren) about the future, which could then have led by association to the nuclear question. The question however was deployed in such a way as to divert her and my attention from worries or fears about the issues the grandchildren would have to face.

Earlier in the conversation, I had asked Charlotte about her images of the natural world, after we had spoken about climate change and other subjects.

C: (Long intake of breath) That's a tricky one. One feels one should do more. (pause)

G: (Elaborating the question)

C: I can live with myself. I know what I've done, I know what I've been. I know I (pause)

G: Yes

C: You've got to live with yourself.....

G: But I wondered if there are particular bits of nature that you think of

C: I don't like the forests being burnt down for the sake of earthly gold or whatever. That I do object to. Um. I fear that we could do more, and I think science could do more to do with the African problem of no water...and to do with the problem of these camps – but they're genocide, a lot of them...
You think, My God, what would you do? And, to be perfectly honest, I love Egypt, and I look at it at the moment, and think, God, you know, what's that - I love what they did at the time it happened to civilization, 2000 years you sit there and you look and think, without any modern.... I sit in awe of that, I must admit. I think a lot – I think out of sight, out of mind. Out of sight out of mind.

G: Is that the case with climate change as well do you think?

C: I jump up and down about homelessness.

This part of the conversation feels very much like the answer to an accusation. There is a fending off of guilt; return to 'we could do more', a correction to 'science' could do more, reference to 'camps' in Africa, a diversion to Egypt, and surprisingly, a conclusion apparently returning to the difficulties in Africa, or perhaps a reference to climate change 'out of sight out of mind'. When I ask directly about climate change, the subject is ignored and the answer comes directly back about homelessness, where Charlotte does feel she has made a real contribution.

A difficulty in researching environmental topics—indeed in having any conversations about environmental issues—is the penumbra of blame which hangs around it. People become defensive even without a hint of accusation (and there are many accusatory lines of discourse associated with environmental issues). This is not to condemn moral discourse about the environment, but to recognize that imputed blame inhibits genuine communication. In a research context, the evocation of this judgemental frame leads to more defensiveness than one might otherwise encounter, sometimes enlightening but potentially also inhibiting respondents' account of their thoughts, feelings and stories.

In some circumstances, respondents took a more light-hearted approach. Jack remarked:

> OK, we may well be contributing to global warming and speeding the process up then, but is that always a bad thing? Y'know. And I would think it's incredibly difficult to say well if we continue doing this, in 3,000 years..... In 3,000 years the world may have disappeared and humans might be living on Mars.

Flippancy, humour and cynicism are techniques by which a speaker can distance themselves from the emotional impact of information; and to avoid showing, or perhaps even feeling at a conscious level, emotions which are unvalued and perceived to be socially unacceptable, particularly for men. The disconnection from feelings parallels the disconnection from the natural world observed by Jaeger et al. (1993), and is seen to be related to it by Rust and Totton (2012).

Minimizing the impact of climate change was also achieved by comments relating climate change exclusively to Africa, or other distant parts of the world. A respondent reported another councillor as exclaiming 'I'm not interested in the bloody polar bears and what have you.'

Diffusing Responsibility

Fairly typical, as among the general population, were statements which placed carbon reduction in the UK in the context of much larger countries whose climate impact far exceeded our own.

> You've only got to look at America, who are, you know, two fingers up to the rest of the world, pardon the expression. But they don't seem to take it as seriously as other countries do. (Amber, council officer)

Similarly, when asked about the relevance of individual actions, Richard a council officer, made a rather despairing comment about a lot of little things adding up to a little, and added:

> There's no point in the West saving carbon emissions, if China and India are increasing by a bigger amount.

Diffusing responsibility can also occur in a more immediate and local sense, for example in bemoaning the already-mentioned short termism in local government.

> I don't think we discuss long term issues enough. And I don't know whether that's because we don't have an answer – we can't even begin – I think there are probably some councillors can't even begin to see – I think most councillors struggle to see a life beyond their next election date. (Nicola, councillor)

'We' becomes 'some councillors', and then 'most councillors'. The issue of whether we can see an answer at all is glossed over. It may be more tolerable to blame other councillors for short-sighted self-interest than to admit helplessness.

Being 'Realistic' or 'Practical'

One councillor, recounting that she had taken part in anti-nuclear power demonstrations when she was a student, laughed hectically as if it were an amusing early aberration. Laughter often seemed to indicate embarrassment at approaching the border to socially acceptable attitudes, in this case taking a passionate stance on environmental issues seemed to be beyond the pale (though one can be passionate about food poverty or homelessness).

I have already quoted Charlotte, who counterposed worry or concern about the future to her own 'realistic' attitude. Being 'realistic' or 'practical' is often contrasted with being 'green'.

> And to be fair in this case as well as any others I'm sure, there are some issues about green – fine – but actually in the real world..you know. –Jack, councillor

> I'm not green, like some people who sit in the council, because I'm also practical. –Grace, councillor

> I wouldn't say I am terribly green. What I'm trying to say is that I view it as more common sense than – trying to see all sides to it. –Charlotte, councillor

> They talk like this, they think 'Oh no, She's green' kind of thing. I do all these things. –Beatrice, officer

This attitude can be associated with impatience with, or scorn for the idealistic stance of those who advocated more vigorous action on climate change. One respondent positioned himself as 'hard-headed' in relation to me, whom he saw as an 'environmentalist'.

When Charlotte broke off from a consideration of the future for her grandchildren, to describe herself as realistic, I felt not only that she was avoiding a subject which could have evoked anxiety, but that, in contrast to her, I was being identified with an idealistic position which would not gain traction in the real world.

> One thinks of their....but I tend to be realistic. How realistic is it – I don't know - you can only go with your common sense. I hope I've got some common sense. What are your views on nuclear then?

The request for my views on 'nuclear' seems especially relevant in terms of Charlotte positioning herself in opposition to a view she would have regarded as idealistic or 'green'. Hoggett identifies such attitudes as 'pseudo-realism', embodying a discourse which draws on social Darwinism, and adopting a stance which is disdainful and superior in relation to another who is seen as naïve and unrealistic. The effect of this, Hoggett observes, is to 'discount the possibility of constructive human agency. While it passes itself off as realism it is actually fatalism' (2012).

Agency

Although most respondents agreed with the statement that 'our organization can play a part in responding to climate change', conversation revealed that many, if not most, are deeply pessimistic about the chances of realistic mitigation measures. They tend to the view that even if the UK meets its emissions targets, other countries will not reduce their own emissions, and that carbon reduction hampers businesses without achieving any impact at a global level.

Furthermore the sense of agency in local government to tackle problems of any kind has suffered as a result of budget cuts from 2011 on, and the reduction of their power over development through new planning guidelines.

> It makes people say, well it's somebody else's agenda, and they'll do what they like anyway. –Morris, councillor

David Ballard (2005) comments that pro-environmental changes to promote sustainable development require three conditions: Awareness, Association and Agency. Ballard, reflecting on a six-month training he carried out with senior departmental managers for a company, considers that Awareness might have to follow rather than precede the other two conditions. He observed that only when they were already charged with their task, had a real sense of agency, and had developed sufficient associative relationships and safety within the study group, were these managers open to taking in more fully the findings and implications of climate science. He concluded that it might not be possible even to take in full information about climate change unless both this sense of efficacy and the strong and supportive group were already in place.

Pam Lunn compares the range of responses to climate change, including denial that it is occurring, with documented attitudes towards Jews in Nazi-occupied Europe. Rescuers tended to have a greater sense of efficacy than bystanders, as well as being likely to hold strong moral commitments which led them to feel that 'anyone would do the same', whereas 'bystanders' who did not engage in rescuing felt weak and alone: 'What could I do against so many?' (Lunn 2011).

If people declare that they are powerless, is this a mask for self-interested reluctance to use what little power they have, in this case securing reductions in greenhouse gas emissions? Or is the denial and avoidance of thinking about climate change documented in the literature, actually pain relief for those powerless in the grip of a hegemonic political and economic system heading towards destruction? A struggle with guilt and responsibility evident in some of the responses quoted above suggests that the first explanation may have some relevance, whereas for individuals like Kate (see p. 183), who identifies with her marginalized constituents and battles against the local power structures, fatalism may serve as anaesthetic.

Different kinds of agency are evident within the sample. Gordon, a senior councillor maintained a long-term batting average despite others in his party and opposition from senior officers, by proposing, asking questions, requesting reviews and maintaining strong links with environmental and Transition groups in his constituency.

On the other hand in Weirbridge, Alison, a senior officer, had hired a cinema, arranged a showing of the *Inconvenient Truth* (with popcorn) for officers and members; had talked about climate change as she placed objects in the time capsule to be buried under new development, and over a long period of time circulated information to councillors about climate change, and identified relevant conferences and trainings to which members were 'sent'. Clearly she was not the only senior person who contributed to bringing this agenda to the fore, but these actions may have helped to legitimize discourse about climate change and to create the sense of association which enabled information to be absorbed and the conversations about policy to be taken forward.

Attitudes to the Future

Respondents' were invited to comment on the card statement 'I am contented and confident that those who are now young will inherit a good future'. Only three of the sample agreed with the statement; 23 disagreed (Table 10.2).

Most respondents mentioned a variety of concerns, including besides climate change, unemployment, the risk of world conflict, food and energy prices and living standards.

> I think we're living in probably the best era I think (laughs) Well, I don't seem to have struggled financially or in every other shape or form, I think we've had the easiest time. –Miles, councillor

> I think our generation have robbed our younger generation. I had free education, mandatory grants, good house prices. –John, councillor

Others referred more explicitly to climate change. Comments in this group included

> I'm very hopeful. I would like to be completely confident, yes. I would like to be. … One feels very challenged at the moment, particularly with jobs and austerity and…which seems to have taken precedence over the climate change and those sorts of issues I feel in our public debate at the moment, and one worries, one worries a great deal, you know…. it just feels to me like we're on a slippery slope. You can tell I'm a pessimist'. – Bernard, officer (twisting and crumpling cards in his hand)

Table 10.2   Outlook on the future.
Q: I am contented and confident that those who are now young will inherit a good future

	Agreed with the statement	Don't know or question avoided	Disagreed with statement
Agreed there is not much an individual can do about climate change	2	2	5
Disagreed there is not much an individual can do about climate change	1	2	18

Contented and confident....... Rubbish. No, I'm sorry, I'm very, very pessimistic about my grandchildren and great grandchildren's future.....
I believe immense change is possible, if the political will is there, and I don't see the political will. –Gerard, councillor

I so want to believe that! (laughs). Whether I do or not, I'm not sure. – Annabel, officer

This group which answered pessimistically with the statement 'There is not much that one individual can usefully do about climate change', paradoxically, tended to be more active within their sphere. Respondents in this group were also willing to admit to more intense feelings about the issue. Table 8.2 indicates that of the 23 who did not feel confident about a good future, 18 thought that individuals (and therefore that they themselves) have some agency. Here again, one might hypothesize that a sense of agency allows the tolerance of more difficult information and feelings about prospects for people in the future.

Conclusions

Attitudes found in this study give support to theories of denial elaborated by Norgaard, Stoll Kleemann and Hoggett. The interview data provides many examples of techniques and social conventions of discourse which distance the respondents from the issue of climate change and from responsibility in relation to it. A close reading of the interview material with a psychosocial perspective reveals some strong evidence of what Rowson (2013) calls 'stealth denial' and Cohen (2001) 'interpretive' and 'implicatory' denial, acknowledging in some degree the existence, and in most cases the human contribution to climate change, but splitting off any feelings about it. Interviews provide brief glimpses of guilt, sadness or anxiety which transgress socially sanctioned norms and might occasion defences.

Intriguingly, they also show an absence of climate change scepticism when it comes to providing flood defences or maintaining communication networks against the impact of climate change. Naturally, agency is not lacking in relation to adaptive measures, and the benefits are local and tangible, much less uncertain than the benefits from reducing emissions. This suggests an element of material self interest (classically defined as 'rationality') in what people and institutions choose to ignore.

Respondents' comments highlight how the circumstances of local authorities in England today, experiencing around 40% cuts in central funding, have much magnified financial and other short term worries, which crowd out consideration of longer term and wider issues and erode any sense of agency in relation to them.

Socially approved repertoires appear, including constant reference to financial advantage, which are linked with disregard either for climate change as an urgent problem or for the possibility of action to combat it. Statements are made which, I have argued, are better understood as strategies for maintaining a positive self image or justifying action or lack of it, than as expressions of belief or ignorance. Bearing in mind Weintrobe's analysis that these are ways of managing affect, we should not be too surprised, therefore, to find that the underlying anxieties are not fundamentally assuaged, and emerge even more strongly when respondents are talking about the future which their children or grandchildren's generation will inherit.

Paradoxically, it seemed to be respondents with a greater sense of agency and of the value of personal action, who more frequently expressed pessimism about the future. Further studies might indicate whether as David Ballard suggests, a sense of agency is needed for full awareness, or whether some other quality or commitment explains both. It is clear, however, that neither better information about climate science, nor recognition of the emotional responses which fuel denial are sufficient to remove obstacles to action. This requires also the experience of working together with others (association) and a realistic sense of agency for both individuals and corporate bodies such as local authorities. Such a sense of agency would exist in the presence of clear (and probably financial) incentives and policies for emissions reduction which create confidence that widespread action will be taken, and with promulgation of more accurate information on action currently being taken to mitigate climate change in other countries, cities and local administrations.

References

Ballard, D. (2005). Using learning processes to promote change for sustainable development. *Action Research* (Special Issue on *Change for Sustainable Development*), *3* (2). http://alexanderballard.at.zilla.org.uk/wp-content/uploads/2012/05/Using-Learning-Processes-to-Promote-Change-for-SD-final.pdf.

Cohen, S. (2001). *States of denial: Knowing about atrocities and suffering*. Cambridge: Polity.

Hoggett, P. (2012). *The psychosocial perspective*. Paper presented at 'Psychosocial Dimensions of Climate Change' Seminar, University College London Energy Institute.

Jaeger, C., Dürrenger, G., Hastenhoz, H., & Truffer, B. (1993). Determinants of environmental action with regard to climate change. *Climatic Change, 23*(3), 193–211.

Lunn, P. (2011). *Costing not less than everything: Sustainability and spirituality in challenging times* (Swarthmore Lecture 2011). London: Quaker Books.

Norgaard, K. M. (2006). 'People want to protect themselves a little bit': Emotions, denial, and social movement nonparticipation. *Sociological Enquiry, 76*(3), 372–396.

Norgaard, K. M. (2011). *Living in denial: Climate change, emotions, and everyday life*. Cambridge and London: MIT Press.

Rowson, J. (2013). *A new agenda on climate change: Facing up to stealth denial and winding down on fossil fuels*. London: Royal Society of the Arts, Action and Research Centre. Available from: http://www.thersa.org/action-research-centre/learning,-cognition-and-creativity/social-brain/reports/a-new-agenda-on-climate-change. Accessed 16 Jan 2014.

Rust, M.-J., & Totton, N. (2012). *Vital signs: Psychological responses to ecological crises*. London: Karnac.

Stoll-Kleemann, S., O'Riordan, T., & Jaeger, C. C. (2001). The psychology of denial concerning climate mitigation measures: Evidence from Swiss focus groups. *Global Environmental Change, 11,* 107–117.

Weintrobe, S. (2013). The difficult problem of anxiety in thinking about climate change. In S. Weintrobe (Ed.), *Engaging with climate change: Psychoanalytic and interdisciplinary perspectives*. London and New York: Routledge.

Westcott, G. M. (2017). *The role of subjective factors in local authorities' action on climate change in South West England*. Ph.D. thesis, University of the West of England. Available from: http://eprints.uwe.ac.uk/28966.

Zerubavel, E. (2006). *The elephant in the room: Silence and denial in everyday life*. New York: Oxford University Press.

11

We Have to Talk About...Climate Change

Robert Tollemache

Introducing Denial

This chapter outlines research based on twenty interviews conducted in 2012–2013, typically lasting between sixty and ninety minutes, with twenty mostly middle-class Londoners. My aim was to understand people's different motives as to why they resist action to reduce fossil fuels use and CO_2 emissions. The majority of my respondents were aware of and concerned about climate change, but they denied in their own interestingly different ways the need to take personal responsibility to mitigate climate change and environmental damage, and defended their non-environmental behaviour. All my participants had complex and contradictory thoughts and feelings about climate change. For many, climate change and environmental concerns were low on their order of priorities. It could contribute substantially to mitigating climate change if people talked more freely about it instead of avoiding the subject, but this ignores the powerful social pressures not to talk about 'difficult' subjects.

R. Tollemache (✉)
London, UK

© The Author(s) 2019
P. Hoggett (ed.), *Climate Psychology*, Studies in the Psychosocial,
https://doi.org/10.1007/978-3-030-11741-2_11

Childhood relationships had shaped my participants' thoughts and feelings about climate change and the environment. Those committed to environmental concerns had often been close as a child to an environmentally committed parent, and felt attached to the countryside of their childhood. Those brought up in a city with close and supportive parents often had no feeling for the natural environment.

As an experienced psychotherapist with a lifelong interest in overseas development, I hoped to apply qualitative social research methodology from a psychoanalytic perspective to explore people's unconcern about ever-increasing atmospheric CO_2. I drew on the many quantitative studies of resistance to taking action on climate change, but I sought a more individual and deeper understanding. Using a modified psychosocial research method, analogous to a psychotherapeutic approach but without interpretations, I tried to help my participants to establish their own narrative, while also maintaining a focus on their feelings and views about climate change and the environment.

I made contact with participants through friends, colleagues and acquaintances. They all responded knowing that the focus of the interview would be climate change, and were to that extent self-selected. Though not a randomised sample, I aimed for a spread of approximately half male and half female, and half and half over and under age 40. More by luck than design I had two non-Caucasian participants, one from Africa, and one from South and Southeast Asian parents. My participants would mainly see themselves as middle class.

I refer to 'interpretive communities' at various points hereafter. Norgaard's (2011) research emphasised the pressures to conform to social norms that prevent awareness and discussion of the effects of climate change, and Lorenzoni et al. (2007) described how their participants belonged to and were influenced by interpretive or cultural communities.

Denial, Internal and Psychological: Negation and Disavowal

Of course, it is impossible to see denial outside its social context, however denial can be seen as largely or completely internal and psychological. Freud described denial, the phenomenon of holding two opposite positions at once, using the concepts of *Verleugnung* (disavowal) (Freud 1938, p. 204) and *Verneinung* (negation) (Freud 1925, pp. 235–239).

When a repressed fact becomes conscious, but its psychological implications are denied, it is negated. Here is an example:

> we fly much too much; erm, that is, I don't know what that is, that is being spoiled, having the money to do it and doing it.

Disavowal consists in simultaneously disavowing and acknowledging something. Disavowal is reminiscent of 'doublethink' (part of 'newspeak') in Orwell's *1984*, '(T)o know and not to know, to be conscious of complete truthfulness while telling carefully constructed lies, to hold simultaneously two opinions which cancelled out, knowing them to be contradictory and believing in both of them' (Orwell 2013, pp. 40–41).

I illustrate negation and disavowal from my interviews with Francis, an academic, and Holly, a school governor and retired wife of a banker. They both sanctioned their non-environmental behaviour by negating its emotional and ethical impact, enabling them to continue non-environmental behaviours that were important aspects of their ways of life. Francis, a highly educated and able man, was vaguely aware of the detrimental effects of climate change on developing countries:

> I have a... I suppose a vague awareness that there will be certain countries – Bangladesh, parts of Africa – that will be particularly hard hit by climate change.

Holly expressed her open ambivalence by admitting that her dislike of 'negativity' might disguise an acceptance of unwelcome facts.

> I don't like negativity, I don't like people moaning and groaning, I haven't got much patience, actually ... I don't like doom and gloom predictions, although I may think, *arggh* [a noise suggesting fear or anxiety], there is probably some truth in what they are saying.

Francis, a frequent visitor to the USA, disavowed the vitriolic controversy about climate change there, a good example of turning a blind eye to what is actually happening.

I do not remember anybody ever discussing climate change in Texas; indeed I cannot remember any discussion in America, even in (the) kind of national press.

The psychoanalyst John Steiner (1985) revisited the Oedipal Complex to illustrate disavowal. He argued that Oedipus himself, Jocasta, and the Elders of the city, were simultaneously aware of Oedipus's relation to his mother Jocasta, but they all denied it. The sociologist Stanley Cohen described three different forms of denial of the violations of human rights under the South African apartheid regime (Cohen 2001). Francis's unawareness of climate change controversy in the USA was like Oedipus's denial of marrying his mother as Sophocles described it, or Cohen's category of literal denial, denial of the facts.

Holly's disavowal was reminiscent of the pattern where people don't feel concerned about environmental damage unless they have direct experience of it.

I don't know that you can really internalise it until you try it yourself, erm; until you care about something that is going horribly wrong – if you, if you see seabirds covered in oil on a beach somewhere near where you live, all of a sudden it means something to you personally.

Forms of Negation

Some participants in my research did not deny climate change, but saw other issues as more important.

Whilst I think that, erm, the environment affects those kinds of people that my work at [a social enterprise] targeted, it often feels to me like the environment is one of a long list of problems that affect them, erm, erm, and, whilst there might be a way to address some of those individual problems, erm, I am, in my work I am most interested in, erm, increasing their resilience to whatever problems like this come along.

Some participants described knowingly acting in environmentally damaging ways, sometimes clearly feeling entitled to.

we love travelling, jump on a plane, you know, off we go, we do it a lot ... We went to India at Christmas, because we can [he laughs]; we did. Let alone what that means: what that nine hours of plane in the sky there and nine hours back meant for the environment – we didn't consider.

Some were open and honest about their own or other people's reluctance to accept climate change data. Holly said:

I don't like doom and gloom predictions, although I may think, *arggh*, there is probably some truth in what they are saying.

Both activists and sceptics suspected that their opponents were receiving unfair financial backing. Climate change sceptics believed that government funding for research was directed to scientists avowing man-made climate change; activists suspected the fossil fuel industry of providing financial support for climate change scepticism. As one sceptic put it,

The governmental subventions they (climate sceptic scientists) got to support research etc. was refused, effectively, if they were researching anything other than, yes, carbon emissions by human activity.

Forms of Disavowal

Victor, an architecture teacher, was perfectly conscious of thoughts that he then successfully disregarded:

it is a bit like eating meat: if I do it, recognising full well that, how it is produced is often horrendous, and I just choose not to think too much about that, even though, you know, in my head, and through reading and understanding, you know, I know that it is not, not a good scene.

Charles, a retired medical researcher, disavowed guilt feelings and consciously created a split within himself:

I don't feel guilty, no I don't feel guilty. I mean I, I, I sort of, I, I put up a kind of, I put up a wall.

Francis was able to disregard the moral implications of his actions,

> I have, for example, at times in my life flown a great deal, which I under-
> stand to be more environmentally damaging than driving a car, erm; and,
> erm, the thing that I feel probably ought to have changed my attitude – I
> can't say has had a huge effect on my attitude.

A majority of my participants felt despairing about possible action on
climate change. Rosemary, a retired ecological financier, had previously
felt capable of acting but now felt despairing. Like Rosemary, many
of my participants expressed feeling unable to act in the face of forces
beyond their control.

> I don't think the individual, even in a, you know, a relatively sized
> group, you don't make that much of a dent to anything, it has got to be
> worldwide.

Rosemary described her excitement after university, engaging in environmental work and climate change action, when everything seemed
to fall into place for her. She could not recall exactly when her brother's death in a road accident had happened, possibly a disavowal of its
impact. Her response to his death (she threw herself into activity) may
have distracted her from her own real needs and so eventually dimin-
ished her contribution to action on climate change:

> I missed my buddy, but my own life was suddenly kind of really opening
> out, so, the pain that it might have been, was still there, but there was a
> lot of distraction.

Some participants were preoccupied with catastrophes, in two cases perhaps associated with the premature death of a close relative. This might
be a disavowal of the slower long-term implications of climate change.
Some suggested that a major crisis would have to occur before people
were persuaded of the necessity to act on climate change.

> I suppose, I don't know, along with, I gather it's a significant segment of
> the population, I enjoy, or think I enjoy, disaster scenarios.

Karen, a supermarket checkout worker, vividly expressed this catastrophism:

> *Karen*: 'the Arctic is ruined,, the cap is melting left right and centre, and I remember watching, erm...third Day after tomorrow...'
> *Interviewer*: 'how did you see that actually?'
> *Karen*: 'Oh god, I loved that film; I can't afford it, mostly, but, I have seen it so many times, but, you know it is...'
> *Interviewer*: 'you have watched it a lot of times?'
> *Karen*: 'Oh god yes; yes; I love disaster films, I am... I know everything to do, I know all the countries to go to if anything terrible happens, I am telling you, I am well planned.'

Three of my respondents described other people as settling for tokenistic action, questioning environmentally friendly action because it is inconvenient, or becoming numbed. One saw the outlook of country dwellers as conformist and limited. These may have been projections of their own feelings.

Science can be reframed as debate. Man-made climate change is often reframed as a debate between two equally valid points of view. Holly generally accepted climate science, although she was attracted to news items that negated it.

> I did hear something quite interesting on the radio, probably two years ago, which was saying that actually there is just a long-term up and down in the, in the temper–, in the temperature of the earth and that we are simply going through a, a swing one way at the moment and we should not think that therefore that it is, erm, that it's a crisis at all.

Science can be disregarded. Benjamin, a consultant to social enterprises, knew the statistics of climate change but believed that people's needs in developing countries today override the needs of future generations, especially since it may be too late to achieve change. There can also be ignorance of science. Two of my respondents may either not have known that greenhouse gases like CO_2 cause long-term climatic change, or they were in literal denial, where knowledge of the facts is denied. Science can also be contradicted. A minority of participants questioned climate science and disregarded the scientific consensus of man-made climate change. Others, like Clive an FE teacher, were frustrated and despairing that since scientists disagree, people ignore environmental problems, and will not act to prevent catastrophic climate change.

Empowerment and Disempowerment

How far participants felt able to take action influenced their responses to climate change mitigation. Some from secure, stable and supportive family environments, expressed optimistic and empowered attitudes, and had pro-environmental views and concerns about climate change. Pamela, a theatre worker, had close, loving relationships with her grandparents, though her mother died in her infancy and her father was mentally ill. Arthur, had a clear sense of purpose as an environmental journalist. Their optimism helped them to feel empowered to act on climate change, saying that they could not carry on if they were not optimistic. According to Arthur,

> You read about the fact that, countries like Greenland and other Arctic countries actually see it as an opportunity then to extract oil from places that were cut off. It is a tragedy, and it is, it is, erm, yeah, I mean it is a shame, it is really depressing sometimes; but, erm, you have just got to be, you do have to be optimistic that we can, we can create change, erm, and, it probably will have to come in the 11th hour really.

Gramsci wrote 'I'm a pessimist because of intelligence, but an optimist because of will' (Marqusee 2012).

Empowerment and Entitlement

People's optimism was sometimes associated with their feeling entitled. Holly felt entitled to fly, describing her frequent flying and heli-skiing as 'wicked'; but she did it because she could afford to. Pamela clearly had misgivings and felt she was hypocritical, but still flew for weekend city breaks:

I do like, you know, going away for the weekend to somewhere that I need a short-haul flight but, if it is, if it is Paris or Brussels, I would always take the Eurostar even if it – and I do mean that – even if it cost me more money.

Some self-confident respondents held orthodox free-market positions, they seemed more certain than the optimistic, environmentally oriented and self-questioning participants.

Edward, a city solicitor, subscribed wholeheartedly to market values and wealth creation. A climate change sceptic, he represented himself as independent-minded, but conceded that there was considerable climate change scepticism within his social circle. He was unconcerned about extremes of wealth and poverty, with a *reductio ad absurdum* argument that it was possible to imagine a situation where the poorest 25% of the population might be multi-millionaires with six cars:

I discovered the other day, that they (the Rowntree Trust), when they talk about poor people, they mean the bottom 25%. So the bottom 25% could be multi-millionaires, with six cars in the garage, but they would still be poor because they were in the bottom 25%.

Francis saw his academic colleagues as having simplistic views because of their limited experience of the world outside the university. He explicitly advocated market mechanisms to deal with distributing resources, though he accepted that they may allow the rich to do what others cannot do, for instance, travel, including flying. He preferred to risk damage to future generations rather than ask people today to sacrifice their living standards:

you're never going to stop people flying or driving cars, erm, and therefore there has to be a decision as to who is going to do it; and, erm, the reason why I think markets are useful are because they are transparent, so there is no messing around about what is being done, erm, and they are efficient.

Edward and Francis denied any social pressures from their social circles, but they each revealed ways in which they were perhaps influenced by the interpretive communities to which they belonged. Benjamin's economics background and commitment to poverty reduction led him to dismiss climate change mitigation on behalf of future generations, given the need to fight poverty in today's world.

> I think I just go by the rule of thumb that, erm, having done some economics, intergenerational thinking is very challenging ... So, if you were to try it ov'.. over a long enough timescale, then, you know, maybe the industrial production pretty much just needs to stop, erm; so when you say that you want to look after future generations, erm, the question becomes like, 'Well, which ones?'

Francis and Benjamin accepted the facts of climate change but dismissed their importance. Edward seemed empowered by his promotion of climate change scepticism, and Benjamin by working for poor people in developing countries. In their study of values, Schwartz and Bardi (2001, pp. 21–24) proposed polar opposite value clusters they termed self-transcendence and self-enhancement. Self-enhancement has widely been interpreted as in line with individualist values, self-transcendence as similar to egalitarianism. Edward and Francis implicitly subscribed to self-enhancing values. Benjamin subscribed to a combination of individualistic and egalitarian values. This combination of self-assurance with received economic thinking seems associated with literal denial, as with Edward, or implicatory denial, following Cohen's analysis, given how Francis and Benjamin acknowledged man-made climate change but disregarded its implications.

Disempowerment, Pessimism, Passivity and Helplessness

A majority of participants expressed pessimism and varying degrees of disempowerment. Some saw declining public interest in climate change (interviews were conducted between late 2012 and summer 2013 a few years after the fiasco of the UN Climate Change Summit in Copenhagen), some experienced increasing damage to the environment and some were explicitly pessimistic about mitigating climate change. A number hoped that forces outside themselves would achieve action on climate change, and many of these felt alienated from the social groups or interpretive communities they belonged to.

Gloria, a comparatively young NGO executive, and Victor thought about the environment more when it was covered more in the media, and believed that changed political and economic circumstances were responsible for the reduced coverage. Gloria felt that this had influenced her own declining interest:

> (I)f I constantly see articles about climate change, er, it, it just keeps reminding me that it is obviously an issue, and, erm, and, yeah, I think that the fact that there hasn't been very much this year,... So, yeah, maybe people have just lost a bit of enthusiasm. I mean, I certainly feel slightly less enthusiastic about the whole thing (*Yeah*) this year.

Victor:

> it is not, doesn't seem to be on the agenda any more, you know there was a time in the late Nineties, early 2000s when it very much was, to my mind; and I thought about it more then, probably.

Several participants experienced a broader sense of decline encompassing climate change, the environment, society and even in themselves. Mary, an arts and the environment worker originally from the USA, had a narrative of environmental decline which was based on her view of her father's adoption of industrial and chemical farming processes, which had degraded the Midwestern farming environment where she grew up:

> I remember when I was a child, the cats on our farm... were very healthy; by the time I left they all had weepy eye problems and all kinds of things, and one can only think that that is on a farm, much more about the water they drink and, and the environment they are living in.

Victor looked back nostalgically to a world in which many people took being self-sufficient for granted. His wife's parents are still like this.

He thought earlier generations who were more self-sufficient were similar to environmentalists today, and imagined an ideal world in which consumption would become less important:

> The difference between our generation – not so much my parents' generation but probably the generation before theirs, the grandparents' generation... they grew their own food and they, erm, had a slower pace, arguably...there wasn't the same come consumer pressures to update, ... they were self-sustainable but in a different way, because that is what you did.

Again thinking of his childhood, Arthur described the feeling for countryside being swallowed up by urban development:

> The town was expanding, so you would see certain parts of the countryside disappear,... you want to, kind of, sound the alarm for, for the places that you, you have seen wild. You know, even as a five year old kid, you know that a rabbit can't live underneath a road, a tarmac road, so – and that is quite affecting.

Rosemary recalled a time when she knew what was important and what needed to be done regarding climate change. Now, since the destruction of the World Trade Center and the financial crash, she felt much less certain about anything:

> in the Nineties it was like, 'Yeah, we know what we are doing, right; we have got a plan'; erm, and, and now I am like, you know, the combination of that – the Twin Towers was the first whammy, and then the crash was the second one – the crash revealed, all of us who had been working in finance, for having absolutely [she laughs] no clothes on; because none of us had focused on the totality of finance.

Some participants were pessimistic about enough being done to avert environmental catastrophe. Albert veered between feeling pessimistic about anything effective being done and then avowing a hope that, after all, things could change. Charles described being very unhopeful, and saw himself as different from his social world in his concern about climate change:

> I don't think many people think as much about it as I, you know, go round, I mean – no, I think I am thought of as being a nerd.

Some participants, like Louis, looked to events and actions outside themselves to prevent catastrophic climate change:

> Oh yeah, I mean, I am; I am disempowered; I mean, it is a, I mean, empowerment is political, it is basically politics…. apart from my own personal energy to do stuff – but, if I become part of a group, then, then there is more power there.

Social Sources of Denial

I will now turn to how participants see the influence of their families, their peers, their social worlds, and the view of society they hold in relation to climate change and the environment.

Family Influences

A close relationship with an environmentally minded mother, who modelled environmentally friendly behaviour, and a childhood attachment to a particular rural environment, disposed people towards pro-environmental thoughts and feelings. Arthur was brought up in a small town on the edge of the country. His mother involved him in her interest in the natural world during their daily walks with their dog:

> I did always have an interest in, in wildlife, nature, erm, which I probably attribute to the fact that my mum was very good at, you know, we always went walking the dog, for example – every day, erm, rain or shine, erm, over fields, so naturally I had a connection with the countryside. …
> in five minutes, well, you know, two minutes you would be in a field,

and that does make a big difference I think to the way you appreciate, erm, wildlife.

Rosemary was brought up in the country by a mother who was passionately involved in the countryside.

> my mum in particular, erm, was passionate about all things outdoors; she was a fanatical fisherwoman, erm. We grew up with loads of dogs around, and ponies when we were little...I grew up as a tomboy, outside, birdwatching, mapping out where all the nests were, on the ponies, going off exploring.

The form of denial that each participant adopted was often connected to their family of origin. Francis's father and wife had scientific backgrounds, and his own university colleagues inclined him towards scepticism about climate change:

> I can remember, er, my father being rather contemptuous of, er, kind of early green arguments.

Benjamin was brought up in an inner-city environment. His father was brought up in rural West Africa and struggled to get educated given his own father's opposition. Benjamin had no interest in the natural world, perhaps because of his inner-city upbringing:

> Yeah; [inner-city London] was an environment I was comfortable in... if you kind of looked to our shared experiences, say, like the odd school trip, a lot of them would be firmly within, quite an urban context. So ... talking about the environment often, can often feel like a, quite a nebulous concept to me.

Simon, a stage technician, was an only child and was born and brought up in east London. He had a close, supportive relationship with his mother, who encouraged his interest in the theatre in order to build his confidence. He felt threatened by and avoided the country, and returned to an urban environment as quickly as possible.

In London you are, you are never alone, you can feel lonely – incredibly lonely – but you are never actually alone, whereas in the countryside it is just you. You are left with your thoughts.

Participants' disavowal was sometimes expressed by projecting onto the wider society their own personal attitudes and disregard of the science of climate change and its implications for future generations. Francis felt detached and unconcerned about the fate of future generations beyond a certain point in time:

There must come a point at which generations in the future become increasingly abstract to you, and you don't, er, you don't care about them in the same way.

The Influence of Social Groups

Only one of my respondents, Edward, campaigned wholeheartedly for an unequivocally sceptical position, providing a platform for the infamous sceptic Professor Ian Plimer within his professional milleu. In contrast, although most of my respondents were environmentally aware and took small steps towards engaging with these issues only a few had ever been involved in openly campaigning and none were committed enough to make major steps such as the avoidance of flying.

Some of my respondents professed strong self-transcendent values and associated with people with similar values but even so their engagement with climate change was limited. Brian felt part of a like-minded social circle:

I would say that most people in my social network are on the same page as I am… people accept climate change; they accept it's anthropogenic; they accept that it is escalating, and that we need to do something about it … We are, we are all in much the same space.

He felt strongly influenced by the idealistic, younger fellow students on his recent development studies course, which perhaps reflected his own self-transcending idealism. Nevertheless he felt a disconnection between these new values he'd acquired and his behaviour:

> *Brian*: 'Going and doing this course, that was a huge discovery. Because there you are surrounded by mainly academic literature; and be… you know, a group of 22- to 35-year-olds, who would be more passionately committed to the whole environmental and ecological argument than I was on entry.'
>
> *Interviewer*: 'Are you regarded as strange among your friends and relations?'
>
> *Brian*: 'I think I'm not. No, I don't think I am seen as strange…. What it means is that I'm not doing, I'm not living, I'm not living the life, I'm not walking the talk. I'm talking it but I'm not walking it'.

Gloria felt a conflict between her humanitarian, environmental interests and the more hedonistic activities of her friends. The pressure to conform to her social group demonstrates how an interpretive community (Norgaard 2011) influences its members. She could be seen as manifesting conflict between self-transcendent and self-enhancement values (Schwartz and Bardi 2001):

> I was into the same things as they were, you know [laughs] – sport and drinking and all that, erm – … I'd always shown a bit more interest in the world, like, you know, like reading the news and taking part in different demonstrations and things, and, no, I mean, I just sort of got my friends to come along if they wanted to, I mean I, I don't know.

Victor felt that he belonged to a middle-class group who subscribed to mostly environmental values, existing in a sort of bubble. He would conform to a decision by society to act in environmentally friendly ways, but would not act individually, or take a lead:

> Most people in my generation are very good at recycling – actually not in my generation – but we then are, sort of [he laughs], a bubble within the society.

Arthur felt that he had a circle of friends who broadly subscribed to the same environmental values as him, but again there were limits to how far these values were manifest in action:

> My friends definitely have an awareness of the environment… they would go, erm, and maybe, do a bit of volunteering, or, erm, you know, go and, go and watch whales, knowing that they are contributing to, erm, safe-guarding of species… and they make ethical decisions about where they go on holiday, but ultimately they, they, it is very difficult to get there ethically, to get there in a way that isn't going to contribute to, erm, a warming climate.

Arguably, many participants like these above manifested the influence of their interpretive communities (Norgaard 2011; Lorenzoni et al. 2007). This was also manifest in the stories told by the few sceptics in my sample and by those, such as Benjamin, who simply didn't see it as a pressing priority compared to other social justice issues. Benjamin worked in social enterprises and had altruistic values, but subscribed to traditional economic beliefs, including the overriding importance of economic growth. Benjamin's friends shared his views, and agreed with his distinction between environmental and economic perspectives. His social grouping seemed to conform to Lorenzoni's description of interpretive communities.

> I think most people have, in my circle have, some level of concern about it; it is very, it is often focused in quite limited areas, so, like, erm, so most people I know would do recycling, or things on those lines… I think a lot of the people, say, who studied development with me or my, kind of, friends more generally, they are often, they often take quite an economic perspective on things, where they are concerned about inequality in the world.

Many participants distanced themselves from deeper engagement by disparaging or ridiculing environmental activists. Gloria avoided 'lecturing' her friends in order not to be thought of as an environmental extremist.

> It's all about eating out, doing all these different fun things, erm, which, you know, you sort of disregard what kind of impact you might be having, and I, it's just not fun to be thinking about it all the time, either, erm; and I just don't want to lecture any of my friends or anybody.

Like several other respondents she did not want to be seen as 'preaching'. Committed environmentalists, including those who foreswear flying, draw attention to conflicts that others feel, so people distance themselves; they disparage or ridicule those who challenge social norms with a committed environmental stance. This recalls Zerubavel's (2006) account of the sanctions that are applied to silence breakers. Several respondents described committed environmentalists as crazy, hysterics, misanthropic, fundamentalist or judgmental. Their reactions could be what Marris describes (1982) as arising from interruptions in structures of meaning that cause anxiety and bewilderment. Pamela described how militants can be perceived as crazy, and David saw activism as leading to woolly thinking, thus ruling out any sort of activism. Edward portrayed environmentalists as alarmist, charlatans, subconsciously hating humanity and behaving discreditably. Louis wanted to avoid the fundamentalist judgmental stance he saw in other people.

Holly and Pamela expressed mixed feelings about leading an environmentally friendly life. Pamela felt strongly about cycling and the way that motorists sometimes treat cyclists:

> I think it is immoral that some people sit in a massive car and they are one person and they are driving, you know, half an hour across London … I am just one person and you have, you're taking up four cyclists' space in a massive car that is polluting the atmosphere.

Committed environmentalists expose the conflict between being environmentally aware and retaining self-enhancing values. Although Holly thought she might be mocked for her environmental stance on recycling she made fun of her environmentally committed sister- and brother-in-law who had given up flying altogether.

On the other hand several respondents admired people that have made a serious commitment to environmental values, including not flying. Despite her mocking of them Holly had been strongly influenced by her husband's sister and her partner. Recycling was a kind of core value for her. Victor implicitly admired his less materialistic, non-flying friends:

> There are one or two notable exceptions that have decided that, 'Right, we are going to be a bit different about this', erm, you know. I think there is one couple in particular, you know, they have decided to do things in a different way; ... They are not income obsessed – they don't, you know, I mean they don't – they are just consuming less. They are not so bothered.

Arthur described the contradiction between his and other people's environmentalism and his need to travel by air for his work, or for holidays.

> my ex-girlfriend's mum, erm, actually wanted to go on holiday to Italy, you know; she was going by train, and that was the only way because she wasn't going to contribute, but most people – you know, she would be one of the exceptions rather than the rule – have a fairly cavalier sort of attitude about it.

Karen the supermarket checkout worker, dealt with middle-class environmentalism at her work via the metaphor of supermarket plastic shopping bags:

> I do not mind doubling up bags now, but, erm, the other bags, they weren't recyclable, they weren't biodegradable, you know, that kind of thing, ... I do kind of joke with them, the customers, you know, like, they are saving the planet, and it is like – 'Yeah, one tree at a time' – you know, that sort of a thing.

Conclusions

A psycho-social approach seeks to ground a person's experience of the world in the context of their life history because our biography contributes enormously to the structure and content of our internal world which powerfully influences the meanings we give to our outer world.

Our experience of the world, including our experience of climate change, is the outcome of a constant two-way dialectic between inner and outer realities. My research indicates that for many of my interviewees the meanings they gave to nature, the countryside, and climate change were explicitly influenced by their families and their social identities, and the social groups to which they belonged. One might wonder whether, had I been able to investigate further, I might not have exposed more family and social influences.

What emerges from this study is the extraordinary power that influences individuals to deny the evidence that stares them in the face about the threat of climate change, and their refusal to engage fully with the problem let alone take much action to achieve change. This was particularly marked in the way many participants dissociate themselves from committed climate change activists, seeing them as fanatical. It is clear, too, that interpretive communities exercise powerful social pressures that discourage talk and action about climate change.

This is particularly remarkable as my respondents were to some extent self-selected as having some interest in climate change by virtue of their responding to my invitation to interview them about it.

Evidently, nearly everybody that I interviewed felt divided and baffled about the problem, and many participants felt helpless, disempowered and despairing; even those who felt empowered and active acknowledged their own inner struggles with feeling helpless.

The single most important factor that could bring about real change is that governments, environmental agencies and concerned individuals make concerted efforts to bring the subject more forcefully to the attention of the general public, and foster focussed and informed discussion of problems associated with climate change.

References

Cohen, S. (2001). *States of denial: Knowing about atrocities and suffering.* Cambridge: Polity.

Freud, S. (1925). *Negation.* SE, XIX, pp. 233–240.

Freud, S. (1938). *An outline of psycho-analysis.* SE, XXIII, pp. 139–208.

Lorenzoni, I., Nicholson-Cole, S., & Whitmarsh, L. (2007). Barriers perceived to engaging with climate change among the UK public and their policy implications. *Global Environmental Change, 17*(3–4), 445–459 (Online). Accessed 29 Oct 2015.

Marqusee, M. (2012). On "pessimism of the intellect, optimism of the will". *ZNet* (online). Available from: https://zcomm.org/znetarticle/on-pessimism-of-the-intellect-optimism-of-the-will-by-mike-marqusee/. Accessed 7 Feb 2016.

Marris, P. (1982). Attachment and society. In C. M. Parkes & J. Stevenson-Hinde (Eds.), *The place of attachment in human behavior* (pp. 185–201). London: Tavistock.

Norgaard, K. (2011). *Living in denial: Climate change, emotions and everyday life*. Cambridge, MA: MIT Press.

Orwell, G. (2013). *1984*. London: Penguin.

Schwartz, S., & Bardi, A. (2001). Value hierarchies across cultures: Taking a similarities perspective. *The Journal of Cross Cultural Psychology, 32*(3), 268–290. Accessed 17 Nov 2015 (Online).

Steiner, J. (1985). Turning a blind eye: The cover up for Oedipus. *International Review of Psycho-Analysis, 12*(2), 161–172. Accessed 12 Sept 2016.

Zerubavel, E. (2006). *The elephant in the room: Silence and denial in everyday life*. Oxford: Oxford University Press.

12

Engaging with Climate Change: Comparing the Cultures of Science and Activism

Rosemary Randall and Paul Hoggett

Introduction

It is well known that most people don't allow news of climate change to trouble them much. To the dismay of those who are more realistically worried, it appears low down on the general population's list of concerns (Capstick et al. 2015, Ipsos Mori, April 2016). Psychotherapists use the term disavowal to describe this process of holding disturbing knowledge in a split-off part of the mind and there is a growing literature which uses psychoanalytic and psycho-social theories of psychological defence to explore aspects of denial about climate change (Lertzman 2015; Norgaard 2011; Randall 2009; Weintrobe 2013). Less explored are the emotional responses of those who do acknowledge the reality

R. Randall (✉)
Rosemary Randall, Cambridge, UK
e-mail: ro@rorandall.org

P. Hoggett
Frenchay Campus, University of the West of England (UWE), Bristol, UK
e-mail: paul.hoggett@uwe.ac.uk

© The Author(s) 2019
P. Hoggett (ed.), *Climate Psychology*, Studies in the Psychosocial,
https://doi.org/10.1007/978-3-030-11741-2_12

of climate change. In a piece of research, conducted between 2013 and 2015 we interviewed a small sample of people who allow its disturbing reality to penetrate their day-to-day lives: climate scientists and climate activists.

Climate scientists confront the evidence daily in their research and have no choice whether or not to think about it, but climate activists and campaigners have chosen, deliberately, to place it at the centre of their lives. The cultures of the two groups are very different. Scientists are located in quasi-public institutional settings whilst activists and campaigners operate through the network-based forms of organisation that characterise social movements. Our expectation was that although a thorough understanding of climate change would in itself be emotionally disturbing the organisational setting in which a group operates could influence the way in which people managed this challenge. We were particularly interested in what supported their emotional resilience because this affects people's capacity to sustain an action in stressful environments (Luthar et al. 2000).

Managing Powerful Emotions

There is strong evidence that an understanding of climate change and its present and future impacts can provoke powerful emotional reactions such as grief and anxiety (Head 2016). Recently, personal and anecdotal accounts registering the distress of those working in the climate science field have begun to appear (Thomas 2014; Kearns 2014) and there is now a website (http://www.isthishowyoufeel.com/) chronicling some of these experiences, particularly those of scientists. Feelings such as grief and anxiety can motivate action but if they are too strong or are accompanied by feelings of powerlessness they can immobilise people or trigger powerful defences.

Two complementary research traditions study how potentially stressful emotional experiences are managed in organised settings. The Tavistock tradition (Armstrong and Rustin 2015) focuses on social defences against anxiety. These are organised and largely unconscious ways in which anxieties, inherent in particular forms of work, find expression in the culture, structure and procedures of institutions, often with negative results as in the high turnover, sickness levels and dissatisfaction which Isobel Menzies Lyth described in her ground-breaking work on nursing (Menzies Lyth 1960). In contrast, Arlie Hochschild (1979, 1983) stresses the more conscious "emotion work" through which emotions, such as anxiety, shape and are shaped in organisations by people taking cues from others and expressing emotion according to rules implicit in the organisational culture.

The two traditions are complementary and combining these perspectives enables us to envisage a continuum of emotional responses. At one pole are intentional and strategic responses, such as the deliberate suppression of pessimism or the open celebration of success, which are part of the "emotion work" of the group or organisation. At the opposite pole, there are unconscious and unreflexive ways of responding, involving organised denial or dissociation. Our expectation was that the flatter, network-based forms of organisation that supported activism would offer a more flexible setting in which to manage some of the disturbing feelings aroused by climate change than the hierarchical and bureaucratic institutions in which scientists work.

Method

The research was conducted using psycho-social methods which are designed to investigate the lived, embodied, less tangible and more affective dimensions of human experience (Hollway and Jefferson 2013; Clarke and Hoggett 2009). Using key contacts in the climate science and activist communities we interviewed six climate scientists and ten climate activists. Interviews lasted for up to 90 minutes, focused on the ethical and emotional challenges that had faced our respondents and situated these in the context of their broader life history. We undertook a "thematic analysis" (Braun and Clarke 2006) of the data in order to let the data "speak" free from our own theoretical preconceptions as much as possible. Following this, the findings from each group of interviews were fed back for further discussion, an opportunity that our activist subgroup seized with much greater enthusiasm than the scientists.

With the exception of one young climate scientist, the rest of our sample were senior figures in the field, four being past or present leaders of research centres. With the exception of one activist in his late 50s, our sample of activists were all in their late 20s, 30s or early 40s. The older ones had become involved in environmental campaigning in the 1990s. The younger ones' involvement dated from around 2005. At the time of the interviews, five of the ten were working for environmental or specifically climate change-based NGOs. Nine of the activists had been involved at some point in direct action and four of them were still closely involved in actions against airport expansion, coal mining, against financial institutions, and against oil companies' involvement in the art world. Extracts from interviews are coded (A for activists, S for scientists) and numbered (1–10 for activists, 1–6 for scientists).

Knowing About Climate Change: The Emotional Impact

The Activist Trajectory

Whilst the majority of our activists could trace a love of the natural environment going back to their childhood they were motivated primarily by social justice values rather than more recognisably environmental ones.

> I don't want to be in a new system where we don't have fossil fuels but we still have racism, and we still have sexism. (A9)

Several expressed a diffuse but strongly felt sense of social responsibility, something they often connected to their family/cultural background, for example, to grandparents who first settled in the UK from India, or to Christian Socialist or Catholic parents.

Virtually all our interviewees described passing through a number of phases of engagement, so much so that we feel justified in referring to the activist trajectory as one that typically involves a rapid awakening to the issue followed by total immersion and then, for some, crisis and burnout, and a subsequent shift towards a more sustainable, proportionate and enduring activism based upon the creative ways of coping that we list later.

Epiphany

For most of the activists the sudden awakening to the issue was an epiphany which involved an adjustment of their sense of what life meant:

> the implication of that (the concept of tipping points) –was what had driven me to put, you know, everything on the line, and you know, stop doing what I was doing and just focus on climate change. (A3)

The findings had a big emotional impact even when returned to after a whilst:

> I always do a couple of days each year updating myself… then I teach the session…and then I need a week of emotional recovery from it because it is absolutely frightening. (A10)

Immersion

This epiphany was followed by a period of complete immersion, reading, thinking, talking and acting:

> It just became the primary thing that I worried about. (A6)

The idea of epiphany also captures the life-changing aspect of the awakening:

> …a friend said 'Oh, there's going to be an action…against… coal-fired power stations and would you…you know, it'll probably be arrestable, but you know, would you like to get involved?' and I just thought yeah… that feels, that would feel like an empowering thing to do. (A2)

Crisis

For all the activists, immersion in the issue involved intense periods of activity which required considerable organisation and often quite high risks. All the time the urgency of climate change was on their mind. Some remembered the impact of the knowledge:

> ...constantly engaging with this stuff, every day...yeah, that was absolutely terrifying... (A3)

For others there was a point when their increasing knowledge had made them feel the problem was possibly insurmountable and they felt overwhelmed, disillusioned or disempowered:

> I did some lecturing at the time on the process of the UN Convention on Climate Change and so I really understood how that part of the UN works and I think it was that process that ended up making me feel very depressed. (A6)

More often people described states such as anger, guilt, self-doubt, confusion and bewilderment, states that are often seen as an important part of the process of dealing with major loss, as the person struggles to remake a sense of what life is about and find new meaning:

> I was constantly anxious... I just couldn't cope...it was an existential crisis...I had to like coach myself to let go of the stupid things that society tells you that have actually no, no bearing. (A1)

And sometimes the problem of how to respond felt almost impossible to resolve:

> I think...lots of us, kind of didn't know where to...where it's OK to draw the line between what you need, what you want for your own well-being and what your duty is as a citizen of the world to do, given that we're in this huge position of privilege. (A2)

Resolution

In their resolution of this process almost all the activists emphasised three forms of emotion work. The first was the development of a sense of agency:

> Just the day to day experience of agency makes the intellectual feeling of despair a bit less. (A8)

> ...despair is a very rational reaction...action is the antidote to despair... (A3)

The second was the need to find a path that was proportionate both to climate change and to living a normal life:

> You definitely kind of have this sense when you're young and you're fresh to it of why are people not...why isn't everybody doing this? And then you realize that well, you can't...you can't stay in that place for ever. (A2)

The third was a conscious moving away from an intense preoccupation with the facts and the pain of knowing them:

> I don't click on it... (A7)

> I barely think about climate change now. It's in the background of my life all the time but I rarely sit and actually talk about climate change or read very much about it. (A6)

> I don't think I've suppressed it...I've accepted it, found my own kind of path of like how I live my life with those kinds of things going through it. (A1)

Although a slight exaggeration there is some truth in Carter's suggestion that "activists have an unwritten agreement never to mention the subject in order to avoid depression" (Carter 2015, p. 577). This distancing felt like the process people go through when they come to terms with living with a difficult condition such as arthritis. After the first stage of finding out everything they possibly can, people settle into getting on with life as best they can. The difficult knowledge is not ignored or denied but is put into the background. For the activists this did not mean that distress did not sometimes return, breaking through the protective barrier:

> I still do sometimes get my climate panic on, you know, thinking about it…yeah, yeah, existential gloom you know… (A3)

> If I let open the floodgates, it's there, I know it. It's not something I enjoy… (A4)

Despite these painful recurrences the activists all seemed to have made healthy and realistic adjustments.

The Scientists' Trajectory

The experiences of the scientists were quite different. For some, the realisation of the implications of climate research came gradually. Typically it was not the findings generated by any single piece of research which were disturbing but the accumulation of findings in connected fields of research combined with the lack of commensurate policy response:

> Yes, I can't think of a point when I said 'Oh my God, got to get into this because this is really important'. (S4)

In general, the scientists were more reserved in talking about the personal impact of scientific knowledge. Where the activists' stories spilled out and they spontaneously shared what they felt, the scientists found our questions about their personal feelings harder to answer. Although they felt deeply concerned about the possible futures which their knowledge opened up, they tended to talk about it in terms of the burden of responsibility which they felt. This strong sense of social responsibility was an ethic which had played a role in drawing several of them into this field in the first place:

I always liked science and this is science with a purpose really essentially and I find it very satisfying. (S1)

But, as the following example indicates, this motive then brings its own problems:

I mean one of the reasons (I went) into this research was because of the value it has for society. But I also…I felt burdened by this responsibility…I had a period of months where I felt very scared because I thought well if the carbon sinks are saturating I'm one of the world experts on carbon sinks. What if I miss it? What if I don't see it? (S2)

Speaking about the so-called "Climategate" affair which erupted alongside the Copenhagen negotiations in 2009, one said:

And the context of course in which that all came out made it that much worse… And you couldn't help them, for they couldn't help but think well we've actually put the boot into that whole process. (S5)

Where the realisation of the implications of climate research occurred more suddenly, the emotional impact seemed to be greater and the effect on personal lives and careers more significant. One researcher temporarily withdrew from the work, another has gone very public about his concerns. A third, recounting how this realisation had offended both his commitment to social justice and his love of "the beauty of this stunning planet," said:

[It's] completely changed how I view my life and what I should have to do in my life… it's coloured my view on almost all aspects of what we do now, in ways that I must admit from a personal perspective has not been positive. It's affected friendships, it's affected family, it's affected the things that I enjoy doing. It's affected my career. (S4)

The scientists were quite varied in how much they had allowed their knowledge of climate change to affect their personal lives. The activists' understanding of the issues had led them to make deep and often difficult decisions about issues such as flying and they all led very low-impact lives. This was also true of some of the scientists but for others the impact was more restricted.

> It's interesting that I did not properly calculate my carbon footprint for years, in spite of the fact I knew I could. I always put it off and of course it was because I didn't want to know. (S1)

Engagement in the Public Sphere: The Emotional Impact

The Activists' Experience

Given our initial expectations what we hadn't anticipated was that for both groups their most difficult experiences arose not from their deepening understanding of climate change but when using this knowledge to engage with a resistant public. The most difficult experiences for the activists concerned police assault, police raids, police infiltration and court hearings. Although all those involved in direct action emphasised the excitement, solidarity and sense of purpose involved, there were also experiences that had been traumatic:

> ..when the police have come round and they've fucking walked on our bed in their boots and emptied all the cupboards and drawers out, to no real effect you know, it's just what they do...that's traumatic and the neighbours seeing the police pouring into the house and things... (A3)

> After we'd done the action and found out that we were going to be defending ourselves in court...I was really lost...I was pretty fragile... (A2)

These experiences had sometimes created psychological damage that was not easily recovered from. One activist who had stepped back from direct action was still distressed many years later as she talked to us. However in the group as a whole, there was recognition of what these experiences were, what their effects were likely to be and what was needed if people were to survive them and recover from them.

The more insidious and less dramatic effects of continual overwork, burnout or depression were mentioned by nine out of our ten activists. The worst experiences had taken place in larger NGOs where in one case the culture of overwork and in another the intransigence of bureaucracy had led the interviewees to leave their jobs. All the other mentions of burnout and depression, however, were countered by discussion of how the interviewees had managed to deal effectively with these experiences and although the people in the two more serious cases had taken longer to recover their stories featured a similar mix of self-care, psychological insight, management of disturbing news and commitment to a different kind of future. The following quote was typical:

> I had... three, really depressing emails in a row about how far it's gone and I just remember thinking that... Oh, my good God, like this is just a lost cause you know, it's a lost cause. But then I have...you know...I talk to a friend...I talked to my neighbour and he said something like 'I just built a wind turbine today, we've just got a community wind project' and 'da-da-da' and I just think great, that's brilliant, that's exciting, so that's what I mean about the solutions keep me going. (A4)

The Scientists' Experience

Although it was not our intention to be selective, we began to realise that our scientists were part of a minority in their profession who were engaged in public communication, in discussions with policy makers and in challenging their colleagues about the responsibilities of climate scientists. This confronted them with the puzzling phenomenon of public indifference and had put them at the forefront of media attacks but they had also experienced conflicts within their own profession which had been very distressing. Discussing public indifference, their strongest emotions were frustration and anger:

> So we're running out of time to be able to avoid the worst risks and...the public debate on the policy process is simply not reflecting the scale and the urgency of the problem. (S3)

However, their deepest distress focused first on battles that developed with colleagues about how the science should be presented to the public and then on the reception that some of them had received in the press when they tried to explain the implications of the science. They described bruising disagreements with colleagues:

> We just become so aggressive with one another in a way that is really not supportive... (S2)

There were occasions where people became fearful of speaking out:

> Climate scientists have now become afraid of any sort of situation in which they feel there is a dispute, and they're trying to avoid that at all costs...they don't want to have a conflict with their opponents and I think that's not helpful to anybody. (S3)

There were accusations and counter-accusations that information had been wrongly presented:

> What we've been saying is challenging to other groups... So we have had a lot of problems I think there and it's mostly been behind the back type comments. (S4)

And there were frightening attacks from the press:

> There are people out there who watch everything that I say...wait and try and find something, and then when they find it they use it to cause trouble. So I feel very threatened and intimidated, and you see it's changing my behaviour. (S1)

Managing the Emotional Impacts: Building a Supportive Activist Culture

In contrast to the scientists, the activists were more easily able to express and discuss what they felt, partly through the active creation of a culture that could face and support the difficulties they encountered. There were three aspects to this: a positive and concrete view of the future, a sophisticated and supportive network of practice, and an emphasis on self-care and proportionality of response.

Ideas of the Future

Despite their pessimism and despair the activists had a clear sense that they were working towards something valuable. Positive ideas of social change offered an anchoring which gave them strength:

> Ultimately, solutions are going to look like more active citizenship…more democratic forms of decision making, decentralised everything really from energy generation to food production. (A5)

A Network of Practice

More important than this idea of the future however was the sense of community that existed now. Trust, support and solidarity were words that recurred frequently when they described their movement:

> There's an incredible sense of solidarity that comes out of doing a direct action. (A1)

> Climate camps were really something special, friends I made then are still friends now. (A10)

Innovative ways of organising which emphasised inclusiveness, consensus decision-making, flat hierarchies, psychological preparation and proper debriefing were mentioned by all our interviewees:

> We build into it after the event doing something where we talk about the emotions of how to deal with that. (A10)

> We have Activist Trauma Support, we have medical support, we have debriefings, we have a really good way of helping people. We know what burnout is now. We know what post-traumatic stress disorder is. (A4)

Self-Care

Individually they were careful to make time for restorative practices, such as meditation, yoga, spending time in nature, maintaining other interests, getting enough sleep, taking holidays and time off, and spending time with family and friends:

> The things that create resilience are just the fairly standard things…just making sure you've got good support around you and having fun when you can. (A6)

Managing the Emotional Impacts: The Use of Institutional Defences

The low-impact, consensual culture which the activists created and participated in was seen by them as a pre-figuring of the future. Although this might sometimes be idealised, it provided a clear way forward at both a personal and political level. Scientific culture, in contrast, creates social defences which seem to have made any working through of anxieties either unnecessary, a source of ambivalence, or a source of continuing pain and difficulty.

The scientists described several characteristics of science and its practice that we thought protected people against anxiety but they also described points where these social defences were breached. The characteristics we identified as performing these defensive functions were ideas of scientific progress, scientific detachment, rationality and specialisation, scientific excitement and normalisation of overwork. These characteristics tended to isolate the scientific work from its implications, cut the scientists off from many of their natural, human responses to their discoveries and allowed the idealisation of science as a neutral good in itself. All the people we interviewed participated to some extent in these defences but had also been involved in challenging them and had suffered as a result.

Scientific Progress and the Excitement of the Work

All the scientists conveyed a strong sense of how exciting the work could be and the older ones had come into the work with a straightforward sense of scientific progress: science was a neutral force whose discoveries would contribute to the greater, human good. However, a sense of inevitable progress can easily lead to the avoidance of the negative consequences of scientific knowledge and to its idealisation. Several of our participants spoke of the moment when for them this assumption had been challenged and some of them saw it as still holding sway amongst colleagues:

> Yes it [the idea of progress] clearly has been challenged and the naïve view if you like that science is just science and do it because it's worth doing and interesting. (S5)

One of them described the conflict of being very excited by a discovery which showed that matters were getting worse:

> I mean part of me was excited when I made the discovery, and oh my God the carbon sinks are saturating, isn't that exciting. Oh my God I should be having the opposite feeling. (S2)

And a number of them described how colleagues would bury themselves in the excitement and rewards of the work, denying that they had any responsibility beyond developing the models or crunching the numbers:

> I know so many scientists that all they want is to do their research and soon as it has some relevance, or policy implications, or a journalist is interested in their research, they are uncomfortable. (S2)

Detachment, Rationality and Specialisation

The idea of scientific progress and the excitement of its tasks are closely linked to ideas of scientific detachment and rationality. Several of the people we interviewed described how colleagues used the ideal of scientific neutrality to distance themselves from the implications of their research:

> ...and many of them will say, oh well it's just my job to chip away at the frontiers of knowledge, and toss my knowledge in and it's up to somebody else what to do with it. (S1)

The scientists described a culture where feelings tend to be distrusted and not acknowledged, where responsibility could be split off and passed to others and where it was easy to become numb to the implications of their work:

> I think a lot of scientists convey the impression that they have no feelings at all about these issues. (S3)

Rationality and logic are clearly second nature to scientists: their operation was apparent in the careful and measured responses of the scientists and the words themselves are repeatedly frequently across all six interviews. In contrast, the word "rational" appears just once in only one of the ten interviews with activists and the word "logic" makes no appearance at all.

The need for specialisation in science made the abdication from responsibility particularly easy:

> ...I mean what scientists do is they − many of them, they pick a bit of it - they pick a bit of the jigsaw and work at it in the hope that other people are working on other bits...and in the end someone, or some way it will be put together and we'll understand the picture. (S5)

Whilst not deliberately designed to avoid anxiety, specialisation nevertheless serves this purpose. One of our respondents was very clear about this. For him it was the process of joining up the pieces that contributed to his mounting sense of anxiety:

> Climate science is very vulnerable because virtually nobody knows firsthand all the pieces of the argument. (S1)

Overwork

Overwork also seemed endemic in the scientific community stemming usually from the pressures of career, competitiveness and fascination with the subject. For our interviewees however overwork seemed to be connected to their urgent sense of responsibility:

> And I think you're right because my reaction to that feeling [anxiety] is to just get on and do the next thing, because that gives me - isn't it a feeling of efficacy, I feel that I've got a grip of things, I'm in control. (S1)

> I don't like switching off. (S4)

In the admission that he didn't like switching off, this last interviewee displayed the defensive function of overwork. Manic activity can defend against depression and the loss of hope: the culture of academic science certainly offered many opportunities for this line of protection. This is perhaps the one defence that both groups had in common for, as we noted earlier, the hyperactivity of many activists also seemed to serve as "the antidote to despair" as one put it.

Managing the Emotional Impacts: Comparing Scientists and Activists

The activists thus had opportunities unavailable to the scientists: the weight of scientists' existing culture did not offer many openings for innovative ways of relating and there were many people around them who would not have been interested, or who would not have seen the point, in developing the kinds of emotional literacy and support systems that characterised the world of the activists. When they spoke of these more difficult experiences, the language and metaphors used by the two groups were striking. Both evoked images of warfare and hunting but where the activists described themselves as protagonists, deliberately thinking about strategy and tactics, the scientists described themselves as the prey.

Figure 12.1 maps the different kinds of defences that the two groups used to manage the traumatic and distressing experiences they had encountered. The informal, networked milieu of the activists offered little opportunity for the social defences endemic in the academic community of the scientists. Nonetheless, when the activists talked of the larger NGOs they had worked in we did get a flavour of social defences operating here as well: "You're not allowed to cry at Greenpeace," one joked, whilst others spoke of a culture of late hours, overwork and bureaucratic systems that stifled both feeling and creativity. Two of the activists were also critical of the movement itself for creating a self-contained bubble that defensively excluded people who were culturally different. In both these phenomena, we saw hints of unconscious defences that coped with anxiety through idealisation of the movement and by setting up rigid boundaries and expectations of what is acceptable. We also caught glimpses of a more individually based unconscious defence in a strand of apocalyptic thinking but in general, the activists' ways of coping used the more reflexive forms of emotion work described by Arlie Hochschild and found on the left-hand side of Fig. 12.1.

Creative and defensive responses to climate anxiety

Fig. 12.1 Individual and social defences

In general, the scientists seemed to have a harder time of it. They were operating in organisations whose cultural norms were ill-adapted to the enormity of their discoveries and which struggled to cope with the pressures placed on them by policy makers and the press. They all felt very responsible: for getting the science right, for communicating it well and for repaying the trust placed in them by the public who funded their work. Although they spoke of getting support from colleagues and their institutions, this was qualitatively different from the support that the activists described. Often the scientists were referring to support for the work they were doing, rather than support for themselves personally. They were more likely than the activists to speak of feeling alone or attacked and to show signs of struggling with experiences that were hard to understand or process. Where the activists openly acknowledged the traumatic edge to many of their experiences and had developed quite sophisticated ways of dealing with this, the scientists were more likely to discuss the traumatic experiences of others, moving attention away from themselves, and they seemed to have fewer resources to cope with their own dark or despairing feelings.

Unintended Consequences of the Use of Social Defences Against Anxiety

Whilst it is clear that social defences such as hyper-rationality and specialisation enable scientists to get on with their work relatively undisturbed by the implications of climate change it is also clear that this approach generates problems. There is a danger that these defences against anxiety eventually break down and anxiety re-emerges, leaving individuals not only defenceless but with the additional burden of shame and personal inadequacy for not managing to maintain a stiff upper lip. Stress and burnout may then follow, and as we saw earlier there is mounting anecdotal evidence of scientists either speaking out about this or dropping out of research altogether.

Even if the defences are successful and anxiety is mitigated, this very success can have unintended consequences. By treating findings as abstracted knowledge without personal meaning climate scientists have been slow to take responsibility for their own carbon footprints, running the risk of being exposed for hypocrisy by the denialist lobby. One research leader candidly reflected on this failure:

> Oh yeah and the other thing, yes, very, very important I think that we ought to change the way we do research so we're sustainable in the research environment, which we're not now because we fly everywhere for conferences and things. (S2)

Secondly, these defences contribute to the resistance of the majority of climate scientists to participating in effective forms of public engagement and intervention in the policy arena, leaving this to a minority who, as we saw in the previous section, are then often attacked by their own colleagues as well as the media. The social defences of logic, reason and careful debate have been of little use to scientists in this public sphere and we could hypothesise that the failure of these defences contributed to internal conflict and disagreement as anxiety could no longer be tolerated. In turn, those that do engage become excessively cautious, which encourages collusion:

> There is a mentality in that group that speaks to policy makers that there are some taboo topics that you cannot talk about. For instance the two degree target on climate change…Well the emissions are going up like this (the interviewee points upwards at a 45 degree angle), so two degrees at the moment seems completely unrealistic. But you're not allowed to say this. (S2)

Similar conclusions about scientific caution have been reached in a recent Breakthrough report (Spratt & Dunlop 2018). Worse still, the minority who are tempted to break the silence run the risk of being seen as whistleblowers by their colleagues. Another research leader suggested that in private some of the most senior figures in the field, including government chief scientists and oil company CEOs, believe that the world is heading for a figure more like six degrees:

I'm sort of deafened, deafened by the silence of most people who work in the area that we work in, in that they will not criticise when there are often evidently very political assumptions that underpin some of the analysis that comes out. (S4)

It seems that the idea of the "socially constructed silence", a concept first introduced by Eviator Zerubavel (2007), applies to some aspects of the interface between climate scientists and policy makers.

Conclusions

Our initial assumption was that detailed knowledge of the risks of climate change would arouse powerful feelings for the two groups in our study—climate scientists and activists. We imagined that this would have had a galvanising effect for them, providing a powerful motivation for action. Our research indicated that the reality was rather more complex. The climate scientists in our sample *had* been motivated to engage the wider public about the risks but they felt themselves to be part of a relatively small minority. They felt that a significant number of their colleagues appeared to remain unaffected and resistant to any action beyond the routines of research. Our analysis suggests that a number of aspects of scientific culture and practice probably provide significant social defences against anxiety for this majority. There is a danger that those scientists that do engage with the public are left emotionally isolated and as a result tend to be overly cautious in their dealings with the media and with policy makers. In contrast, knowledge of the risks of climate change had a sudden and disruptive impact on the lives of nearly all the climate activists in our sample. Far from defending themselves against these impacts the activists had developed an emotionally supportive culture which helped them sustain their commitment over time.

We were also interested to find that for both groups in our research it was not so much knowledge of the risks of climate change per se that was most emotionally challenging but the actions which both groups were then motivated to undertake. For the scientists this included having to deal with anxious policy makers, predatory journalists and rivalrous colleagues; for the activists this mostly involved dealings with the police and justice system. Both groups found this emotionally demanding but whereas activists had been able to integrate "emotion work" into their practice, the culture of science left climate scientists ill-equipped to face these challenges.

Clearly, age and gender may play an important role in the differences we found. The activists tend to be younger and the gender balance more even. But there can be little doubt that the activist community is much more emotionally literate than the scientific one and had found more satisfactory ways of dealing with the psychological difficulties they encountered. Could the scientific community learn anything from the activists about how to manage the emotional impact of their findings?

Chris Rapley has recently called for a culture change in the climate science community so that it becomes more open to public engagement (Rapley 2014). We believe there is a strong case to be made for introducing ways of increasing the emotional resilience of the climate science community, particularly if, given the rise of climate-denying populism, climate science becomes increasingly politicised. The forms of professional supervision and mentoring which are common in nursing and some branches of medicine may be too counter-cultural. It might nevertheless be possible to sail with the prevailing scientific culture by introducing approaches such as action learning (Brockbank and McGill 2004) which provide practitioners with a safe but challenging space to reflect upon the ethical and emotional challenges that they face.

Bibliography

Armstrong, D., & Rustin, M. (2015). *Social defences against anxiety: Explorations in a paradigm.* London: Karnac.

Braun, V., & Clarke, V. (2006). Using thematic analysis in psychology. *Qualitative Research in Psychology, 3*(2), 77–101.

Brockbank, A., & McGill, I. (2004). *The action learning handbook.* London: Routledge.

Capstick, S. B., Demski, C. C., Sposato, R. G., Pidgeon, N. F., Spence, A., & Corner, A. (2015). *Public perceptions of climate change in Britain following the winter 2013/2014 flooding* (Understanding Risk Research Group Working Paper 15–01). Cardiff: Cardiff University.

Carter, C. (2015). Denial and despair? *Environmental Values, 24*(5), 577–580.

Clarke, S., & Hoggett, P. (2009). *Researching beneath the surface.* London: Karnac.

Head, L. (2016). *Hope and grief in the Anthropocene.* London and New York: Routledge.

Hochschild, A. (1979). Emotion work, feeling rules, and social structure. *American Journal of Sociology, 85*(3), 551–575.

Hochschild, A. (1983). *The managed heart: Commercialisation of human feeling.* Berkeley: University of California Press.

Hollway, W., & Jefferson, T. (2013). *Doing qualitative research differently.* London: Sage.

Kearns, F. (2014). *Scientists have feelings too.* read.hipporeads.com/scientists-have-feelings-too. Accessed 16 Mar 2016.

Lertzman, R. (2015). *Environmental melancholia: Psychoanalytic dimensions of engagement.* London: Routledge.

Luthar, S., Cicchetti, D., & Becker, B. (2000). The construct of resilience: A critical evaluation and guidelines for future work. *Child Development, 71*(3), 543–562.

Menzies Lyth, I. (1960). A case study in the functioning of social systems as a defence against anxiety. *Human Relations, 13*(2), 95–121.

Norgaard, K. M. (2011). *Living in denial: Climate change, emotions and everyday life.* Cambridge, MA: MIT Press.

Randall, R. (2009). Loss and climate change: The cost of parallel narratives. *Ecopsychology, 1*(3), 118–129.

Rapley, C. (2014). *Time for change? Climate science reconsidered: The report of the UCL Policy Commission on Communicating Climate Science.* London: University College of London.

Spratt, D. & Dunlop, I. (2018). *What lies beneath: The understatement of existential climate risk.* https://www.breakthroughonline.org.au/whatliesbeneath.

Thomas, M. (2014, October 28). Climate depression is for real: Just ask a scientist. *Grist.*

Weintrobe, S. (2013). *Engaging with climate change: Psychoanalytic and interdisciplinary perspectives.* London: Routledge.

Zerubavel, E. (2007). *The elephant in the room: Silence and denial in everyday life.* Oxford: Oxford University Press.

13

Conclusions

Paul Hoggett

Successive reports by the Intergovernmental Panel on Climate Change (IPCC) refer to the 'business as usual' scenario. This represents slow and incremental progress towards reaching mitigation targets, the outcome of market forces and technological innovation rather than any significant governmental intervention. IPCC reports typically suggest this would lead to increases of global temperatures between 3 and 5 degrees centigrade compared to pre-industrial levels. Even allowing for the likelihood that climate change will not proceed in this incremental and gradualist fashion, this scale of increase would tip us out of the benign climatic conditions of the Holocene which have made civilisation possible (Angus 2016; Lorimer 2017; Hamilton 2017). In other words, to carry on 'as usual' threatens the basis of civilised life. Read this carefully, I am not someone who is prone to alarmism (Hoggett 2011), but this is what the science tells us.

The phrase 'business as usual' has a number of meanings. At the macro level of the global economy it refers to the hegemony of a

P. Hoggett (✉)
Frenchay Campus, University of the West of England (UWE), Bristol, UK
e-mail: paul.hoggett@uwe.ac.uk

© The Author(s) 2019
P. Hoggett (ed.), *Climate Psychology*, Studies in the Psychosocial,
https://doi.org/10.1007/978-3-030-11741-2_13

business model which assumes that growth, indeed accelerating growth, is a self-evident good. Many of the people interviewed in this book seemed to be aware of the madness of this assumption and yet helpless to imagine any alternative. They seemed to illustrate the view that it is easier to think of the end of the world than it is to think of the end of capitalism.

At the meso level, 'business as usual' refers to the dominance of business models within systems of governance in neo-liberal societies. Once the preserve of a different way of organising public goods and services, under neo-liberalism central and local government, public services and voluntary organisations have now been overwhelmed by the discourse of the market. Several of the chapters in this book illuminate some of the impacts of this on those with pro-environmental values who work within such organisations. Here the phrase refers not just to the 'business case' for action (in times of shrinking government budgets every initiative has to first pass the 'business case') but also to organisational rhetorics about what is 'realistic' and 'practical' and by implication 'impractical' and 'idealistic' (see the chapters by Andrews, Westcott and Robison for example). It is by such means that those who might otherwise demand that their organisations 'walk the talk' stay silent in order to avoid being branded alarmist or unrealistic. 'Welcome to the real world' as I once heard a male professor say to a younger, disillusioned, female lecturer.

Finally at the micro level, 'business as usual' refers to what passes as normal in everyday life. The ways in which, despite what we know about imminent climate breakdown, we continue our lives in much the same way as ever. We might make small changes to our diet, make a conscious effort to avoid plastic packaging as much as possible, perhaps even put solar panels on our roof, because all these are by now socially acceptable. But (as some of Tollemache's respondents indicate) to stop flying, become vegan or travel in an electric vehicle still raises eyebrows and runs the risk of being seen as odd, perhaps over serious and possibility fundamentalist.

What we sometimes think of as climate change denial is therefore part of the warp and weft of our everyday social interactions at home and work, part of what we feel we can say and do and of what we feel a pressure not to say or do. It is part of the way in which basically good people end up staying silent or not really saying what they want to say (Zerubavel 2006) and as the last chapter, by Randall and Hoggett, indicates this can happen to climate scientists as easily as to the rest of us.

So, like 'race', climate change is a 'hot' topic for conversation, one permeated by defensiveness, resentment and misunderstanding. It has perhaps been *the* major claim of this book that if we want to start developing conversations about climate change then we need to think carefully about how we do this, and what methods we use. In particular we need to get away from linear-rational models of fact-based argument and persuasion to forms of conversation which engage the imagination, make use of imagery and metaphor, elicit stories and narratives, and give recognition and respect to the inconsistencies, paradoxes and contradictions of our efforts to engage with the ethical challenges posed by climate change.

The title of this book is designed to provoke. Even the worst-case scenario of escalating global social collapse leading to nuclear apocalypse would leave some survivors. But the preservation of the species is not what is at stake. Who knows, by the coming of the 'end times' some wealthy survivors may have bunked off in their 'Tesla craft' to Mars where, no doubt, in centuries to come their offspring might well recreate the same havoc that Earth's original survivors left behind. No, what is at stake is not the survival of the species but the survival of this experiment (and perhaps it is not an exceptional one in this universe) we call humankind as something which has value, meaning and integrity. So, in what may be our end times, as we rearrange our deckchairs and seek the solace of the ensemble's calming tunes, let us think kindly of ourselves and listen to each other with compassion. Only in this way might we manage the troubles to come without panic, violence and indignity.

References

Angus, I. (2016). *Facing the Anthropocene: Fossil capitalism and the crisis of the earth system*. New York: Monthly Review Press.

Hamilton, C. (2017). *Defiant earth: The fate of humans in the Anthropocene.* Cambridge: Polity.

Hoggett, P. (2011). Climate change and the apocalyptic imagination. *Psychoanalysis, Culture and Society, 16*(3), 261–275.

Lorimer, J. (2017). The Anthropo-scene: A guide for the perplexed. *Social Studies of Science, 14*(1), 117–142.

Zerubavel, E. (2006). *The elephant in the room: Silence and denial in everyday life.* New York: Oxford University Press.

Index

A

Activists 21, 23, 115, 134, 151, 154, 155, 165, 221, 233, 240–242, 244–250, 255–257, 259, 260

Affect 12, 13, 23, 28, 38, 87, 107, 119, 131–133, 139, 141, 144–146, 187, 195, 214, 247

Agency 27, 86, 115, 116, 123, 124, 154, 155, 163, 170, 177, 180, 186, 192, 198, 201, 210, 213, 214, 245

Ambivalence 26, 29, 31–36, 54, 72, 87, 96, 137, 219, 252

Anxiety 9, 12, 13, 28, 34–36, 44, 102, 113, 122, 123, 130, 154, 206, 209, 213, 234, 240, 252, 254, 256–259

B

Behaviour change 6, 27, 61

Biography 30, 86, 88, 180

Business as usual 2, 3, 22, 150, 263, 264

C

Carbon conversations 102, 130, 157, 164

Climate change 1–5, 7–10, 12–14, 16, 21–23, 25–27, 29, 31, 33, 37, 41–43, 46, 47, 49, 50, 76, 162–164, 170–172, 195–204, 206, 207, 209–214, 217–231, 239–245, 247, 248, 257–259, 263, 265

Climate science 3, 199, 202, 203, 210, 214, 223, 224, 240, 241, 254, 260

Climate scientists 23, 151, 154, 240, 241, 249, 250, 258–260, 265

Cognition 6, 7, 13, 14, 62, 72, 80, 133, 141

Complex systems 6, 141, 147

Containment 13, 14

Contradictory views 87, 89, 120, 124, 150, 217, 219

Conversational 15, 32, 150, 203, 204

Coping 9, 35, 62, 76, 77, 80, 243, 256

Cultural repertoires 203

D

Data analysis 28, 48, 69, 111, 139

Deep listening 21, 22, 28, 149

Defences 9, 10, 13, 72, 92, 95, 103, 165, 213, 240, 252, 256, 257

Denial 9, 12, 13, 87, 116, 119, 129, 150, 196, 197, 202, 203, 211, 213, 214, 218, 220, 223, 226, 230, 239, 264

Despair 9, 12, 13, 22, 29, 56, 76, 107, 114, 115, 118, 154, 162, 245, 251, 255

Detachment 252, 253

Diffusion of responsibility 150

Dilemma 9, 23, 27, 33, 34, 109, 115

Disavowal 12, 129, 130, 150, 196, 218–222, 231, 239

Discourse 29, 43, 65, 68, 87, 196, 201, 205, 207, 210, 211, 213, 264

Disempowerment 121, 224, 226

Dreams 11, 15, 23, 47, 108, 109, 112, 117, 118, 120, 121, 130–134, 137–141, 143, 145, 147

E

Embodiment 47, 159

Emotion 7, 12, 13, 22, 66, 72, 74–76, 78–81, 87, 90, 102, 103, 143, 203, 241, 245, 256, 260

Emotional labour 90

Emotional reflexivity 151, 154, 155, 159, 160, 169–171

Emotional support 102, 104

Emotion regulation 74–76, 81

Entitlement 9, 224

Environmental melancholia 31, 154

Experience 7, 10–17, 21, 23, 28–30, 32, 33, 35, 61–68, 71–74, 77, 78, 80, 81, 87, 90, 92, 102, 103, 107, 108, 111, 112, 114, 115, 117–119, 121, 122, 131–134, 136, 140, 143–147, 149–151, 154, 156, 163, 165, 167, 171, 179, 180, 182, 183, 185, 189, 191, 192, 214, 220, 225, 241, 245, 248, 249

F

Fatalism 210, 211

Feelings 7, 9, 11–13, 17, 22, 29, 32, 35, 38, 41, 42, 45–47, 49, 50, 52, 65, 77, 78, 80, 99, 101, 102, 107–110, 112, 114–116, 118, 120, 121, 124, 125, 130, 133, 140, 150, 153, 157, 159,

162, 181, 204, 207, 213, 217, 218, 221, 223, 229, 234, 240, 241, 246, 254, 257, 259

Frames 7, 8, 12, 31, 65, 68, 70–72, 74

Free association 42–45, 50, 54

G

Government 2, 34, 36, 38, 67, 78, 123, 150, 195, 197, 208, 210, 221, 258, 264

Group methods 86, 124, 156

H

Hope 10, 29, 30, 51, 61, 81, 104, 114, 115, 130, 137, 156, 159, 162, 163, 169, 205, 209, 228, 254, 255

Hyperobjects 133, 143

I

Identity 7, 33, 72, 77–79, 95, 110, 112, 113, 115, 124, 130

Imaginal 11, 14, 22, 23, 42, 43, 47, 116

Indifference 249

Intellectualisation 76

Interpretive community 232

Interview design 33, 204

Interview methods 45

M

Meaning 6–8, 10–14, 16, 31, 33, 44, 45, 47, 63, 64, 77, 80, 86, 108, 113, 115, 118, 125, 131, 134, 146, 147, 178–180, 189, 191, 203, 234, 244, 258, 263, 265

Meta-object 63, 73

Metaphor 11, 15, 16, 22, 42, 47, 52, 53, 65, 70, 71, 79–81, 108, 120, 136, 235, 255, 265

Mindfulness 72, 114, 149, 158–160, 169

N

Narratives/Stories 15, 16, 21, 22, 33, 43–45, 54–56, 77, 86–96, 98–100, 103, 104, 112, 113, 117, 118, 125, 154, 155, 180, 188–190, 196, 197, 203, 207, 233, 246, 249, 265

Nature 5, 6, 11, 12, 23, 29, 37, 42, 46, 47, 51, 55, 56, 65, 66, 79, 81, 111–114, 117, 130, 133, 136, 137, 140, 146, 149–151, 159, 160, 163, 166, 167, 169, 177–181, 185–192, 201, 206, 229, 252, 254

Negation 150, 218–220

Norms, Organisational 96

Norms, Social 150, 204, 218, 234

O

Optimism 77, 115, 121, 224

P

Personification 22, 42, 46, 47, 52–56

Pessimism 78, 115, 121, 214, 226, 241, 251

Phenomenology 64, 111, 180

Powerlessness 29, 240

Progress (ideology of) 12, 252, 253
Psychoanalysis 9, 131–133, 189
Psycho-social methods 21, 26–28, 30, 149, 218, 241
Psycho-social theory 239

Q

Qualitative research 15, 16, 88, 108

R

Reflexivity 22, 89, 155, 159, 163, 164
Relation to nature 180, 187, 190, 191
    adults 151
    childrens 181, 191
Research methods 26–28, 161
Rhizomatic 23, 132, 135, 139, 140, 145, 147

S

Scepticism 204, 213, 221, 225, 226, 230
Self 2, 4, 7, 8, 16, 34, 51, 65, 66, 72, 75–78, 87, 94, 95, 102, 113, 120, 123, 139, 150, 151, 156, 179, 189, 191, 208, 211, 214, 218, 225–228, 231, 232, 234, 244, 249, 251, 252, 256, 264

Sensing 14, 23, 32, 119, 139, 178, 179, 191
Social defences 11, 22, 151, 240, 252, 256–259
Social dreaming 11, 23, 130–136, 138, 140, 141, 146, 147
Socially constructed silence 259
Suppression 79, 80, 241

T

Thematic analysis 93, 241
Trauma 23, 38, 118, 154, 158, 160, 171, 181, 182, 251

U

Unconscious 11, 12, 28, 38, 42, 44–48, 52, 62–64, 66, 76, 80, 88, 89, 107, 108, 117, 120, 122, 130–132, 139, 180, 181, 189, 240, 241, 256
Unthought known 38, 133, 140, 146

V

Values 21, 22, 33, 37, 61, 62, 66, 67, 72, 80, 94–96, 112, 115, 119, 125, 155, 158, 177, 178, 188, 225, 226, 231–234, 242, 264